Mobile Communication and Low-Skilled Migrants' Acculturation to Cosmopolitan Singapore

Mobile Communication and Low-Skilled Migrants' Acculturation to Cosmopolitan Singapore

Rajiv George Aricat and Rich Ling

LEXINGTON BOOKS
Lanham • Boulder • New York • London

Published by Lexington Books
An imprint of The Rowman & Littlefield Publishing Group, Inc.
4501 Forbes Boulevard, Suite 200, Lanham, Maryland 20706
www.rowman.com

Unit A, Whitacre Mews, 26-34 Stannary Street, London SE11 4AB

British Library Cataloguing in Publication Information Available

Library of Congress Cataloging-in-Publication Data

Names: Aricat, Rajiv George, author. | Ling, Richard Seyler, author.
Title: Mobile communication and low-skilled migrants' acculturation to cosmopolitan Singapore / Rajiv George Aricat and Rich Ling.
Description: Lanham : Lexington Books, [2018] | Includes bibliographical references and index.
Identifiers: LCCN 2018010765 (print) | LCCN 2018007315 (ebook) | ISBN 9781498552516 (Electronic) | ISBN 9781498552509 (cloth : alk. paper)
Subjects: LCSH: Cell phones—Social aspects—Singapore. | Foreign workers, South Asian—Social networks—Singapore. | Unskilled labor—Social networks—Singapore. | South Asians—Social networks—Singapore. | South Asians—Cultural assimiliation—Singapore.
Classification: LCC HE9715.S55 (print) | LCC HE9715.S55 A75 2018 (ebook) | DDC 303.48/33095957—dc23
LC record available at https://lccn.loc.gov/2018010765

The paper used in this publication meets the minimum requirements of American National Standard for Information Sciences—Permanence of Paper for Printed Library Materials, ANSI/NISO Z39.48-1992.

Printed in the United States of America

Contents

Chapter One

Introduction

The mobile phone has become the locus through which migrant workers manage their social links with their families back home and with their compatriots. Take the example of Francis who works at a chemical plant in Singapore. He has not seen his 10-month-old daughter except in photographs. When she was born he was at his workplace, some 3,000 km across the Indian Ocean. He received the news of her birth via his smartphone, and photographs via the same device sometime later. Upon receiving the photos, he shared them with his friends across Singapore; friends who have been with him in thick and thin during his last seven years of work there. His friends congratulated him, first over the phone and then afterward when they met him in person on weekend. It is not new for Francis to be away from home for months on end. He has been a foreign worker since he was 20. Now he is 30. However, at least on occasions such as the birth of his daughter, it would have delighted his family to have seen him. His wife would have felt more relaxed after the labor, and his parents would have cherished the occasion to see him. Things, however, did not fall into place for Francis. Financial constraints stopped him from making a short trip to his home to welcome the newborn, nor did he go home for his sister's wedding four months previous. Francis sends money back to his family and he also stays in touch via his mobile phone.

The situation of Francis illuminates the main themes of this book, namely the use of mobile communication by migrant workers to maintain both their connection with family and friends back home and also to maintain their social network in the country in which they work. As per the latest data available, there are 258 million transnational migrants (3.4% of the world's population) around the globe, compared to 175 million at the beginning of the millennium and 154 million in 1990 (UN, 2017). A large percentage of the immigrant population—defined as people who are in a country other than where they were born for a considerable amount of time—belongs to the low-skilled worker category.

In modern societies, migration is often triggered for economic reasons (e.g., globalization of financial markets). Corporations and technologies have spurred cross border movements of migrants in the recent decades. Remittances from labor migrants have propped up the economies of a number of developing countries in Asia and the Pacific. In 2016, seven of the top ten remittance-receiving countries, in absolute figures, were in Asia (World Bank, 2017). In terms of GDP ratio, two countries in Asia had migrant remittances at a rate of 10% or above: remittances accounted for 30% of GDP of Nepal and 10% of GDP of Philippines[1] (World Bank, 2017).

Migration to the industrial regions is a comparatively recent phenomenon in Asian economies. In the context of South-South migration, Middle East remains a much sought after destination for Asians. Labor migrants to the six countries in the Gulf Cooperation Council, which include Bahrain, Kuwait, Oman, Qatar, Saudi Arabia, and the United Arab Emirates, primarily originate from Asia, including from India, Bangladesh, and Pakistan, as well as from regions such as North and sub-Saharan Africa (IOM, 2018). Other important South-South migration corridors in Asia include Afghanistan-Pakistan (3.4 million), Bangladesh-India (3.2 million), Myanmar-Thailand (1.9 million), Indonesia-Malaysia (1.05 million), Philippines-Malaysia (0.62 million), Bangladesh-Malaysia (0.35 million), India-Singapore (0.34 million), and India-Malaysia (0.13 million) (Philippines Statistics Authority, 2017; UN, 2015a; World Bank, 2017).

Cross-border migration has become a hot topic in many countries. Most of the industrial economies in Asia face the paradoxical challenge of becoming competitive players in the global financial scene in the face of a short supply of talent among their resident population (Ahsan Ullah, 2010; Hewison & Young, 2006). In addition, countries like Singapore lag behind in population growth, as the fertility rates are below replacement level, while coping with an increasing number of elderly people. The immigration policy of Singapore government is, thus, aimed at filling this demographic vacuum in the long term, and ensuring the country's optimal economic performance (Yang, Yang, & Zhan, 2017).

A large percentage of the cross-border migrants are in the low-skilled category. The potential gains for immigrants are not always welcome for the host society, which sees the influx as a threat to jobs and an erosion of national identity. The topic has divided public opinion in regions like North America and Europe, often to the point of bringing nationalist, anti-immigration parties to power. Owing to their lack of political entitlements and opportunities for long-term residence in receiving countries, the low skilled groups have to endure increased adaptation pressures in the host societies. Opinion is divided on how well these groups are to be integrated in the receiving countries. In the wake of recurring migrant crisis in the Mediterranean, the United Nations Special Rapporteur on the Human Rights of Migrants François Crépeau observed:

> Politicians are still not up to the task of telling their populations that we need migrants—doctors and engineers, but we also need low-skilled or unskilled migrants. Germany and the UK have started to look at these issues. But it is not happening in France or Italy or other parts of Europe because then it becomes a quarrel over national identity. We should have a discussion on diversity policies—on who we are and how we see ourselves in 50 or a hundred years. (Naravane, 2013)

As noted by the UN official, the onus is not entirely on the individual migrant worker when it comes to integration. The host countries also need to locate themselves within the frameworks of human rights, reciprocity, and sustainability, while addressing the issues of migrant integration.

Among migrant receiving countries in Southeast Asia other than Singapore, the migration policy of the host government is not often tied with the population policy for the long term. In fact, international migration is narrower in scope to mean mainly the movement of low-skilled workers within this region. In Southeast Asia, the return migration regimes are based on bilateral Memorandums of Understanding (MOUs) between countries. For example, Thailand is home to approximately 2 million Burmese workers, in addition to large number of workers from Laos and Cambodia. Migrant workers in Thailand take up jobs in sectors like fish processing and domestic services for up to 25% (Mon, 2010). Despite the country's heavy reliance on the migrant labor, the recruitment industry is under-regulated. A large percentage of workers are trafficked illegally to the country and, hence, are undocumented. Periodic deportations and a pervasive grip of the shadow economy for migrant labor prevent migrants' long-term residence in the country (Lefevre, 2017). International migration is thus more aligned with labor and economy and less with the population projection for the long term.

Similarly, Malaysia, which has approximately 3 to 4 million international migrants,[2] adopts a strategy of dividing the foreign workforce in terms of its "potential" (Nah, 2012). As a result of this hierarchizing, the skilled workforce has more rights and higher level of integration to the country. The state, along with employers, exercises disciplinary control on the semi-skilled and "unskilled" workers (Nah, 2012). Migrant workers in industries like electronics, agro-food processing, and textiles experience forced labor

(Qiu, 2016). Although free trade has allowed for movement of labor workforce across borders, lack of entitlements and regular exploitation of labor migrants are rampant in countries like Malaysia where regulations are in wanting. The US State Department has repeatedly included Malaysia in Tier 3 watch list for human trafficking (Qiu, 2016).

A wide range of phenomena related to migration has received scholarly attention, including economic benefits for migrant-sending and -receiving countries, changing social structures of the home and host countries, cross-border engagement between migrants and non-migrants, and issues of migrant adaptation in receiving countries (Sniderman, Hagendoorn, & Prior, 2004; Spencer-Rodgers & McGovern, 2002). An emerging field of research that cuts across these themes is the role of communication technology in the migration process, cultural adaptation, and cross-border activities among migrants. Information Communication Technology (ICT) in general and mobile phones in particular have become central in an analysis on migration infrastructure, migratory movements, and cross-cultural interaction. While a macro perspective helps analyze the role of ICTs in the migration process (Broeders & Hampshire, 2013), a micro perspective, focusing on individuals and groups, has received wide research attention. This latter perspective is followed in this book.

MIGRANTS' COMMUNICATION PRACTICES

The communication behavior of migrants, with individuals and groups as units of analysis, has received scholarly interest for many decades. Prior to the advent of ICT, this research area broadly included two aspects: (i) interpersonal communication of immigrants; and (ii) mass media usage. Interpersonal communication involves immigrants' interaction with dominant cultures in the host society, which is found to have a positive relationship with their adaptation. In the case of Hispanic and Asian groups who immigrated to the US in the second half of the twentieth century, this entailed their day-to-day interactions with Afro American and European communities already rooted in the country. Migrants also engage with members of their own ethnic community in the host country seeking social support and availing information needed for their survival and adaptation. Low-skilled migrants, who are otherwise segregated from the mainstream host society, have stronger interpersonal communication within ethnic groups.[3]

Mass media use of immigrants is the second area of interest. Immigrants normally use two types of media, which are the English language national media, and the ethnic media. National media carry news related to the host society in general and, thus, give less coverage to the local news on immigrants. Ethnic language media, on the other hand, are published or broadcasted locally in the host countries, but cover news from immigrants' home countries as well. Studies have focused on a number of factors ranging from demographics to media functions to understand the impact of ethnic media on immigrants' adaptation (Fathi, 1973; Hwang & He, 1999; Viswanath & Arora, 2000).

The diffusion of low-cost communication technologies like the mobile phone has provided a fresh impetus to the research in this area. Statistics show a steady increase in mobile phone diffusion around the world, including to the deprived population in the developing regions. At the time of this writing, there are 8.1 billion mobile phone subscriptions in the world, the preponderance of which are in the Asia-Pacific region (GSMA, 2016). Research data also suggest increasing uptake of information and communication technologies (ICTs) by in-country as well as international migrants. Research has also examined the beneficial ways in which the migrants use the technology (Hamel, 2009).

This book focuses on the use of mobile phone by international migrants in Singapore. A unique point of the analysis is that it focuses on low- and semi-skilled male migrants from South Asia—mainly Bangladesh and India—in Singapore. There have been several studies of female work-migrants to Singapore who are usually house help and maids (Chib et al., 2014; Thomas & Lim, 2011; Yeoh, Huang, & Gonzalez III, 1999). However, less attention has been paid to the male migrants who work on construction sites and other industrial sites. The work focuses on the low skilled male migrants' use of the device in their attempt to acculturate to the host country.

The ratio of foreign workforce to the resident population is high in Singapore: the strength of foreign workforce is 1.37 million, as against 3.97 million residents. Apart from the foreign workforce, other categories of non-residents include international students and long-term visitors (0.27 million). Thus, out of a total population of 5.61 million, 1.65 million (29% of the total population) are foreigners. Out of the total population, nearly 45% are foreign born, with most of them originating from Malaysia (44%), followed by China (18%), India (6%), Indonesia (6%), and other countries (26%) (UN, 2015b; Yang, Yang, & Zhan, 2017). Among the foreign workforce, the low-skilled male workers in labor-intensive sectors such as construction and marine constitute a large percentage (0.3 million or 22% of the total foreign workforce) (MoM, 2017a). Workers in these low-skilled jobs are only given temporary contracts, which precludes any possibility for long-term residence or citizenship in Singapore. Their temporary resident status and low economic standing has resulted in a kind of marginalized existence in the cosmopolitan city.

Using the framework of acculturation, this book explores how mobile communication has mediated their acculturation to the host society.

ACCULTURATION AND LOW-SKILLED MIGRANTS

The standard definition of acculturation encapsulates those "phenomena which result when groups of individuals having different cultures come into continuous firsthand contact" (Redfield, Linton, & Herskovits, 1936, p. 149). Since the emergence of this concept, there have been several approaches that empirically validated how this phenomenon occurs in real life (Barry, 2001; Cai & Rodríguez, 1997; Ward, 2001). One approach is to consider acculturation as simply a derivative of the amount of time migrants live in the new culture (Gordon, 1964; Kim, 1989). This is in line with early assimilation research in the context of US, which observed that immigrants adopt American values with the passage of time. A second approach is to follow a more pluralistic interpretation, which argued that immigrants showed behavioral, cognitive, and emotional affinity to both the cultures they are acquainted with—the home and the host (Berry, 1997; Horenczyk & Munayer, 2007).

A major drawback in these two approaches is that they consider acculturation as a theoretical framework that is only relevant for the long-term resident population. As the volume of low-skilled migrants crossing international borders has now increased, receiving countries are also under pressure to integrate these temporary workers in the host society. Workers, on their part, also face adaptation challenges like language difficulties, lack of emotional support, and proneness to exploitation in the host country (Vedder & Virta, 2005; Yeh & Inose, 2003). Thus, acculturation becomes a broader concept that summarizes the array of problems immigrants face in the new culture as well as their response to the problems.

Factors such as immigration policies of the host government, the host society's attitude towards immigration and ethnicity, and class differences within the migrant population have impact on acculturation. Studies have highlighted how these factors act differently on migrant groups, thereby influencing their adaptation strategies in the host country (Ford, 2011; Teo & Piper, 2009). The interactions between migrants and the host society have appeared in various shades of mutual acceptance, tolerance, and hostility (Murray & Marx, 2013; Raijman, 2013; Stephan et al., 2005). Hence, a study on acculturation must take into account the broader context of host society's attitude towards the immigrant population. In Singapore, the response to the problem of immigration has been mixed, especially when it comes to the intake of low-skilled population.

CIVIL SOCIETY AND LOW-SKILLED MIGRANT POPULATION IN SINGAPORE

Ever since colonial times, there has been a steady in-bound stream of bonded and migrant laborers from China and India. This means that Singapore is a multi-ethnic society (Lockhart, 2010; Sandhu, 1993). To accommodate the diversity of its population, the state policy of Singapore has drawn on multiculturalism. This is evident in policies related to ethnic quota system for electoral constituencies and allotment of Housing Development Board flats. The quotas are based on the official categories of Chinese (74.3%), Malay (13.4%), Indian (9%), and other ethnicities (3.2%) (Sing Stat, 2017). Government's immigration policy, which feeds into the population and economic policies, is also aimed at retaining the proportion of races in the long run (Jones, 2012).

Singapore's economic policy aimed at attracting foreign investments between 1960 and 1990 led to the development of industrial estates and export-oriented manufacturing in the country. This caused a spike in the intake of low-skilled migrants in the country. Increased supply of workforce also resulted in decreased stability of jobs. The government's policy has been to repatriate the temporary workforce at times of economic downturn. The recession in 1974–1975 resulted in the repatriation of significant numbers of guest workers from labor-intensive manufacturing industries. After recovery in 1978, recruitment gained momentum, extending its reach beyond Malaysia, which had been considered the traditional source. Currently the guest workers are sourced from India, Bangladesh, Sri Lanka, Philippines, and Thailand (Coe & Kelly, 2000).

The government is cognizant of the potential for sectarian strife and has developed a range of policies to counteract this possibility. In general, the state policies have ensured peaceful and harmonious coexistence between ethnic groups. However, there has been growing concern among the resident population on government plans for allowing sustained immigration to the country. Civil society's apprehensions range from the feeling of economic insecurity and the overstraining of public infrastructure to feeling of cultural dilution:

> Foreigners feed our sense of insecurity. If that sounds like a poor reason for our mistrust of foreigners, we can lay claim to the second: Foreigners, through no fault of their own, cannot partake in a strong and pulsating Singapore identity. (Chang, 2010)

These concerns were reflected in the civil society's opposition towards a White Paper on population promulgated by the government in early 2013. A roadmap to increase the population to 6.9 million by 2030, the White Paper proposed to open gates to 15,000 to 25,000 immigrants every year for the next one-and-a-half decades (Population White Paper, 2013). However, the idea of encouraging migration at such a high rate was subsequently rejected by the resident population (Goh, Yong, & Sim, 2013).

As low-skilled workforce constitutes a large percentage of the immigrant population in the country (MoM, 2017a), there is also an increasing intolerance among the residents towards the immigration of this class of workers. Low-skilled and low-income migrants from certain nationalities suffer greater discrimination in social spaces in Singapore (Chan, 2011; Jacobs, 2012). Bangladeshi are stereotyped for their lack of language skills and laggardness, whereas Chinese are considered unruly and "country bumpkins." Although not a regular happening, hostilities towards these migrants find expression in online and mediated spaces, including stereotyping of migrant workers based on nationality (Chia, 2014).

Retaining guest workers as a repatriable workforce is one of the key principles of the government's economic policies. In this context, the relationship between the core resident Singaporean population and the guest workers is that of a customer and service provider (Chin, 2008). It is argued that in such a situation, there is no room for relationships that are enduring. The cultural meanings, symbols, and language that the temporary workers share with the resident population are, thus, superseded by the economic logic of contract relationship. This presents a unique situation in which acculturation can only be understood beyond the similarities and differences in cultural aspects, by factoring in the socioeconomic standing and the human capital of the outgroups.

CULTURAL SIMILARITIES AND DIFFERENCES

State-supported multiculturalism in Singapore has helped resident migrant groups to freely practice their religion, and to organize cultural festivals and events in public places. Cultural centers and public spaces for ethnic gathering like Little India, Arab Street, and Chinatown are well known in Singapore. New immigrants from Southeast and South Asia and China can identify themselves with a mixture of cultural symbols established by their forebears in the host country. These can potentially facilitate their acculturation (Cheng & Lee, 2009; Ford, 2011). Studies have shown that in cultural contexts where segregation between ethnicities is less pronounced, for example the ethnic Chinese in Vancouver or the multicultural fabric in Singapore, new immigrants will enjoy optimal adaptation outcomes (Jurcik et al., 2013; Ryder,

Alden, & Paulhus, 2000). In such cases, similarities between host and immigrant races and cultures facilitate greater assimilation of new immigrants. However, in the case of low-skilled guest workers the underlying dynamics of acculturation seem to be different.

Behind the similarities of overt cultural symbols, there are differences in cultural values and ideologies between the same races that are settled in Singapore and that of the new immigrants. For example, Chua (2003) observes that the racially defined cultural characteristics of the resident population have been eroded in the wake of "westernization" (Chua, 2003, p. 61). Although considered a lingua franca capable of bridging the racial differences, English has been incapable of bridging the class differences in the country. Rubdy and McKay (2013) observe how a "dominant standard language ideology" among Singaporeans disfavors the migrant workers, whose lack of proficiency in English has been evoked as a reason for keeping them in the margins (p. 162). Goh (2013) observes that cultural events and festivities in Singapore have become Debordian spectacles with hybridization of ethnic cultures as their central objectives:

> The dance [performed during the Racial Harmony Day carnival] could be interpreted as a critique of modernity, sending traditions to their graves and then desecrating the graves, in the process replacing traditional Malay matriarchy with patriarchy. (Goh, 2013, p. 233)

For the new immigrants, the similarities of cultural symbols, festivities, and artefacts between the host and home countries can facilitate the acculturation process. Yet, the class differences between the immigrants and the resident population undercut the similarities in cultural practices. The differences surface during immigrants' extended period of stay in the host country.

MIGRANT CLASS DIFFERENCES AND ACCULTURATION

Compared to the professional group of migrants, low-skilled migrants are disadvantaged in terms of economic and status entitlements. This group of migrants remains a transient workforce in the country, which makes it easier for the government to repatriate them at times of economic slowdown (Ahsan Ullah, 2010; Leong & Berry, 2010). Driven by an economic logic, migrant-receiving countries generally deny equal opportunities to migrants across professions. This plays through in limiting their ability to settle in and to assimilate to the host country over the long term. In Singapore, state policies mean that migrant workers remain as low- and semi-skilled work categories despite long years of service and familiarity with the work and social systems

(Kaur, 2010). On the other hand, professionals in sectors such as finance and information technology have more incentive to integrate themselves, even though they only constitute one-fourth of the total workforce of Singapore (0.37 million or 26.9% of the total foreign workforce) (MoM, 2017a). The differentiation between "foreign talent" and "guest workers" in Singapore is rooted in a neoliberal economic logic (Munck, Schierup, & Wise, 2013). In short, it is more a policy of containment and not a policy of integration that has been followed in Singapore in the case of migrant workers (Rahman & Kiong, 2013).

The policies of containment have resulted in a paradox of acculturation for low-skilled migrants in Singapore, reflected in their "long-term but perpetually temporary residency" (Thompson, 2009, p. 360). This potentially affects the migrants' sense of belongingness in the host country, their work performance, and socio-psychological well-being. The increasing discontent among low-skilled population has resulted in occasional outbursts in the form of riots and strikes (Siddique, 2012; "Trouble in," 2013). On the one hand, the civil society in Singapore, especially in the aftermath of incidents like the Little India riots in 2013, urges the foreign workers to "abide by the rules of the land" as long as they are here. While on the other, the resident population is hesitant to consider them as partners or co-participants in the great project of Singaporean development. This has opened up a kind of paradox for the migrant workers in Singapore, wherein acculturation is achieved by continual divide, and distance maintained between the host society and the workers.

It has been shown that ethnic mass media (Christiansen, 2004; Croucher, Oommen, & Steele, 2009; Jeffres, 2000; Reis, 2010) and migrants' interpersonal communication with the host society (Kim, 1978; Schwartz et al., 2010) help bridge divides between migrants and the host society. Ethnic media have helped migrants learn the local custom and assimilate to the new culture (Zubrzycki, 1958). Concepts like cultural hybridization (Hall, 1996), biculturalism, and marginal man (Park, 1928) have long been promoted by ethnic media as progressive ideas that can help society advance in the path of multiculturalism. It is thus important to explore how mobile phones, like the mass media in yesteryears, help migrants acculturate to the host society.

ACCULTURATION AND MOBILE PHONES

Mobile phones are being increasingly adopted by migrants at the low-income strata. The last two decades saw a surge in the number of studies on migrant workers' use of mobile phones, suggesting a mobile phone revolution among the workers (e.g., Law & Chu, 2008; Thompson, 2009; Wallis, 2008). The statistics on the diffusion of mobile technologies among the migrant population around the world are not comprehensive. However, studies have noted the rapid diffusion of the mobile phone in the developing world in general (GSMA, 2017; ITU, 2015). Trends are pointing to a rising graph of mobile phone adoption among the low-income migrants.

A comparative statistics on the mobile phone subscriptions in the home and host countries of migrant workers studied here will provide a background for the discussion. India had 1.18 billion mobile subscriptions as of February 2017 ("Telecom subscriber," 2017). That is, there were 93 subscriptions for each 100 people. In reality, lower than 93 out of 100 people will have mobile phones, since there will be duplications caused by dual SIM phones. In addition, a number of people might be using two or more subscriptions. Nonetheless, this serves as a rough benchmark of mobile phone adoption among the population in general. Mobile subscriptions in Bangladesh stood at 130 million (BTRC, 2017), meaning that there were 83 subscriptions per 100 persons. In Singapore, there were 148 subscriptions per 100 persons as of mid-2016. A comparison of GDP per capita of Bangladesh, India, and Singapore would further provide a background. As of 2016, Singapore had a GDP per capita of USD 52,960; for India it was USD 1,700; whereas for Bangladesh, it was USD 1,360 (World Bank, 2016b).

The target population of this analysis—low-skilled migrant workers—may or may not have been users of mobile telephony in their home countries. While their homes in rural Bangladesh and India remain underserved by the mobile service providers,[4] migrants from these countries are active consumers of mobile services in Singapore. In addition, a large percentage of them have adopted smartphones and data subscriptions in the last few years.

One of the primary needs of the newcomers is to access information that would facilitate their induction to the host country. Mobile phones have emerged as a primary source of information, providing an opportunity for the newcomers to tap into various sources of information. Low-skilled migrant groups largely rely on co-ethnic networks in host countries to access information. Mobile phones have increased low-income migrants' opportunities to be connected with these networks (Lin & Sun, 2011; Sibal & Foo, 2016).

Early studies focused on the limited number of channels, perhaps not more than one-on-one voice calling and texting, which were available to migrants to network with their communities. Later, with the advent of social media and microblogging and instant messaging services, migrants expanded their modes of communication, as they adopted newer and more sophisticated platforms that suited their requirements. Recent research explores how a range of communication modes, including transnational storytelling and gaming platforms, helps the migrants to network with their co-ethnics, to communicate across borders and to interact with the host society (Chen, 2014; Cheong, 2015).

At the same time, migrant networks can also have negative consequences such as enforcing ghettoization and the resulting low engagement with the host society (Cohen, 2011; Ryan et al., 2008). This has been explicated as the constraining side of social capital. Highly coherent networks, often supplemented by mobile communication, pose high entry threshold even for new immigrants of the same ethnicity (Grzymala-KazLowska, 2005). A thorough investigation into the effect of mobile communication is thus indispensable for understanding the process of migrant acculturation.

Studies of mobile technologies in the international migration context have concentrated on their ability to provide social connectivity (Qiu, 2009) and social and emotional support (Chib, Wilkin, & Mei Hua, 2013). The acculturation paradigm has rarely been used in such studies. As noted, this work addresses that gap and focuses on the cultural adaptation of low-skilled migrants. There are multiple advantages in using the acculturation paradigm. First, the acculturation paradigm does not consider the need for permanence in migration, thereby bringing temporary low-skilled migrants to the theoretical fold, despite their economic backwardness in the host country. Second, unlike the assimilationist approach, which dominated the studies on intercultural contact in early years, the acculturation paradigm does not preclude migrants' sustained ties with their home/ethnic culture. Finally, the acculturation paradigm helps us understand the effects of mobile telephony. It does not treat acculturation supported by mobile communication as a derived phenomenon or an end goal in itself (e.g., as a way to gain social support or overcome employee control) but as an ongoing phenomenon situated in communication practices.

An added emphasis in our analysis is to examine the acculturation of the low-skilled migrants into both the work as well as the social sector. It is also theorized that migrants' acculturation attitudes would determine their level of adaptation to the host society. Given the nature of the formal work sector in focus here, organizational commitment has been considered a good index of migrants' adaptation. However, the book maintains a balance between work-related and life domain–related acculturation by observing the overlaps between the two domains. It also supports a migrant-oriented interpretation of acculturation, which is non-deterministic and dynamic in its conceptualization.

In sum, with rapid diffusion of mobile technology into the migrant population worldwide (Law & Chu, 2008; Law & Peng, 2007, 2008; Vertovec, 2004), it has become inevitable that the acculturation model be rethought by incorporating the interpersonal and mass communication aspects of this medium. In the case of Singapore, the host countries' communication infrastructure allows the migrant workers to overcome the digital divide that they know so well from their home countries. That said, lack of financial resources, ICT skills, and language barriers still prevent the migrants at the lower end of the economic spectrum to gain access to, and also benefit from, advanced mobile devices (Qureshi, 2012; Warschauer, 2002). This research addresses the issue of migrants' extant communication relations with the home culture, facilitated by low cost international calls, saturated diffusion of low-end phones, and rapid diffusion of smartphones.

In this book, we are interested in understanding the situation of male migrant workers in Singapore and their mobile assisted integration into society. Hence, we ask in this volume:

- *How does acculturation attitude influence adaptation outcomes of migrants in different life domains?*
- *How does mobile phone communication lead to migrant adaptation in different domains of life?* and,
- *How do cultural identity, acceptance of multiculturalism, and mobile calling to home and host cultures affect migrants' adaptation outcomes in the host country?*

OUTLINE OF THE BOOK

The outline of the book is as follows. Chapter 2 provides an overview of labor migration with specific focus on Bangladesh and India. The debate that ICTs have helped reduce the cost of migration has been reviewed at both the institutional and the individual levels. Chapter 3 begins with a review of studies and theoretical positions related to technology. There is a particular focus on the appropriation approach. The chapter then reviews mobile communication studies among migrants. It concludes by presenting the circular model of mobile phone appropriation that was used to guide the research. The model highlights functional uses of the technology as well as the symbolic meanings derived by the users of the technology.

Chapter 4 reviews the idea of acculturation based on a bi-dimensional model. This model characterizes migrants' cultural relations as first, retention of cultural heritage and identity, and second, relating to other cultures. A critique based on diaspora studies that explain acculturation in non-essential terms has also been introduced to describe the qualitative exploration that follows. The chapter closes by outlining the specific research questions and hypotheses that arise from the analysis of studies in technology and acculturation. Chapter 5 describes the mixed methods approach. Results from quantitative statistical tests as well as from qualitative analysis are presented in Chapters 6 and 7. Finally, chapter 8 synthesizes the quantitative and qualitative findings and discusses the theoretical and practical implications of the analysis along with its limitations.

NOTES

1. In Kyrgyzstan, Tajikistan, Armenia, Georgia, Lebanon, West Bank and Gaza, Jordan, and Yemen remittances are also above 10% of GDP.
2. This figure contrasts with the one presented in the government report on international immigration (Department of Statistics Malaysia, 2017). According to this report, the total number of international migrants in the country is approximately 72,000.
3. In Singapore, large weekend congregations of Filipino in Lucky Plaza, South Asians in Little India, and Thai in Golden Mile Complex attest to this point.
4. Diffusion rates of 93% in India and 83% in Bangladesh are recent statistics. When the study was conducted between 2011 and 2013, the diffusion rate was much lower.

Chapter Two

Labor Migration and ICT

The economic reasons behind labor migration are well documented. Neoclassical theories focus on individual costs and benefits and identify cross-border migration as a potential solution for low employability and low wages in capital-poor countries. Political economy approaches such as world systems theory and dependency theory view migration in terms of center-periphery tension. According to these approaches, people migrate due to the structural inequalities in the peripheral areas, that preclude development and thus viable livelihoods for large sections of the population (Massey, 1988). New Economics of Labor Migration theory emphasizes the collective decision of families in the migration process (Stark & Bloom, 1985). This perspective suggests that families decide to send members abroad in order to hedge against any potential loss of income in future, but also to overcome capital constraints on family production activities. Research has also looked at the life chances of migrants in a holistic manner, without emphasizing unduly on the economic aspect (Delgado-Wise, 2014). In order to understand why people migrate, we wish to consider the situation in their home country in this chapter. The main analysis in the following chapters focuses on the situation of the migrants in Singapore. However, it is useful to consider the broader issues that are associated with migration (e.g., factors in the home country that induce cross-border migration, the role of intermediaries in migration, and the role of Information Communication Technology [ICT] in the process).

This chapter focuses on the low-skilled labor migrants from Bangladesh and India, by highlighting the contexts that trigger this kind of migration. It then analyses cost of migration as a major area in which ICT have made an impact. The chapter analyses the impact of ICT at both institutional and individual levels. The chapter then discusses the problems of digital divide among international migrants as a prelude to discussing their acculturation in the following chapters.

LABOR MIGRATION FROM BANGLADESH AND INDIA: MAJOR TRENDS

Labor migration from both Bangladesh and India are triggered by similar reasons, including surplus labor, high population density, limited resources, and a low rate of economic development in the last few decades (Mosse et al., 2002; Sarker, 2016; Siddiqui, 2005). Both countries have a considerable rural population and also low growth in the farm sector. These are some of the factors that "push" people to consider labor migration. Although in terms of remittance, Bangladesh and India do not compare with each other,[1] in terms of the destination of its low-skilled migrants there are similarities. For both countries, up until the 1980s the Middle East had been an important destination. During that time oil prices supported many construction projects in the region. In the early 1990s the Middle East lost its sheen. At that point, Southeast Asia, especially Malaysia and Singapore, became the attractive destinations for these migrants. There are shared cultural and national histories for Bangladesh and India. Indeed, Bangladesh was part of India before the Partition in 1947. This section presents an overview of labor migration from the two countries and the reasons behind the cross-border movements.

Bangladesh

Bangladesh has a total population of 160 million. It has a total workforce of 60.7 million (2013 est.), in which approximately 45% are involved in agriculture (BBS, 2017). The country has 39% of its workforce in the age group of 15–29. Bangladesh has been characterized as a nation of small, subsistence-based farmers, and nearly everyone in the rural areas is involved in agriculture. Large-scale migration from Bangladesh began in 1976, when the cash-rich Middle Eastern countries undertook massive development projects following the oil price hike in the early 1970s. The crisis in the Persian Gulf during the period 1990–1991 prompted a shift in focus to Southeast and East Asia (for recent statistics on Bangladeshi migrants, see table 1 in appendix A).

Bangladeshi temporary migrant population has been categorized into four groups: professional, skilled, semi-skilled, and unskilled. Doctors, engineers, nurses, and teachers are, for example, considered as professionals. Manufacturing or garment workers are considered as skilled; tailor, mason, and so forth are semi-skilled workers; housemaid, cleaner, and laborer are classified as less-skilled. Most of the Bangladeshi workers employed in Singapore are in labor-intensive less-skilled sectors of construction and marine, which afford low levels of pay.

The rural roots of Bangladeshi migrant labor have been noted by a number of studies. A study conducted among 400 Bangladeshi low-skilled workers in Malaysia's capital city and nearby states in 2011 showed that all of them came from rural districts (Sarker, 2016). A survey conducted in Singapore revealed that around 84% of Bangladeshi migrants were of rural origin, while the remaining 16% hailed from small- and medium-sized cities in the country (Rahman, 2017). In the study reported in this volume, the average age of a Bangladeshi migrant was 28 years, whereas the average duration for which a migrant stayed in urban regions for work or residence was just above 13 years.

While migration from Bangladesh has traditionally been dominated by men, recent trends show increasing presence of women. Economic reasons like acute need of money or sudden economic setbacks generally trigger migration among men. Whereas, in the case of women, additional factors like desire to be more independent and the need to get away from unpleasant family situations are also reported. A study conducted about a decade ago found economic reasons for Bangladeshi women's migration to Malaysia in Southeast Asia (Rudnick, 2009). Most of the women lived in villages where no migrants had traveled abroad. But when the women started to receive stories of migration, success, and independence disseminated through local networks, they imagined a different future for themselves: "When combined with the grapevine stories of success and prestige that surrounds the emigration of men, this information inspired many [women] to follow suit" (Rudnick, 2009, p. 103). Thus, the Bangladeshi women in the study migrated to overcome the shackles of traditional society and to be more independent—economically, physically, and emotionally.

Changing social norms in Bangladesh accord high status for migration, which has also become a strong motivating factor for youth to move abroad for work. Rahman (2017) notes that people of *uccho gusthi* (high lineage) traditionally do not work in other people's houses or on their land. But people of both high and low lineages join together in overseas employment. When they cross borders, the social hierarchies become less of a concern. In fact, Bangladeshi society in general has a higher esteem for those who have been employed overseas, regardless of the nature of work the emigrants do in foreign countries. Similarly, different regions in the country vary in the degree to which they send migrants abroad. This is mainly due to the social networks available to the potential migrants.

"Migration networks," in terms of kinship, relatives, and people belonging to the same locality, are strong among Bangladeshi workers. This helps induce a kind of "chain migration" in which workers who have already migrated provide information, help, and support for potential migrants back in villages. Beyond these social ties, migrants also develop symbolic ties based on shared meanings, memories, future expectations, and representations. Bangladeshi migrants identify themselves strongly with their districts. Thus, belonging to the same geographical region facilitates the flow of information and assistance for these workers. The weekend friendship gatherings in Singapore's Little India, for example, are based on friends belonging to the same districts of their home country. In general, an enabling type of social capital is evident in the case of Bangladeshi migrant networks.

In addition to social networks, migrants from Bangladesh also rely on private employment agencies to secure work visas in foreign countries. Recruitment via private channels, however, incurs high cost for the emigrants. Rahman (2017) observes a general rise of economic costs of Bangladeshi migration over the past two decades. Emigrating to Singapore would cost a Bangladeshi worker around US$3,500 prior to 2001, which doubled to become US$7,000 in 2014. Undocumented migration has also existed alongside regular channels. Sarker (2016) notes that there are 0.8 million illegal workers in Malaysia, of which a considerable percentage is Bangladeshi (pp. 128–130). Illegal residency means the migrants are not bound to any job based on their status. Further, just as illegal channels are cheaper for migrants, they are cheaper for employers as well. Thus, while there are risks associated with undocumented residency there are also advantages.

Similar to the problem of illegal migration supported by recruitment agencies are the issues of people smuggling in the Bay of Bengal region and porous borders with neighboring India. A recent UNHCR report says that about 63,000 people have traveled by boat in an irregular and dangerous way in the Bay of Bengal (UNHCR, 2015). Kumar (2015) estimates that most of the boat people are from Bangladesh and Myanmar. Land crossing of Bangladeshi workers and settlers to India has been a politically contentious issue. While the Bangladeshi government terms it as "illegal crossing," nationalist parties in the receiving country brand them as "illegal immigrants" (Gillan, 2002). We need to note that the workers interviewed for this volume were all legally documented individuals.

In sum, Bangladeshi labor immigration to Singapore can be explained by a mixture of theories related to wage differentials, family-supported decision making, and migrant social networks. There is also a substrata of illegal migration and trafficking, largely triggered by poverty and deprivation among Bangladeshis. Despite the existence of enabling social networks, the cost of migration still remains high for a Bangladeshi low-skilled migrant. It involves the money paid to sub-agents or recruiters who provide logistical and administrative assistance for the migrant's journey. Despite the considerable sums paid to the agents, there is often a good chance that migrants will, for example, land up in jobs that are not promised to them at the time of recruitment, or that the jobs they do get provide little opportunity for additional income via, for example, overtime work. It is thus important to explore how the low-skilled migrants negotiate their lives in the host country with limited available resources and social networks to support them. As communication technology, especially mobile phones, has started to play a key role in migrants' negotiation of daily lives, we return to this theme later in the chapter. More pertinent to this book, it is important to explore whether mobile communication has helped migrants' adaptation to the host society.

India

India has a population of 1.27 billion, with a total workforce of 513.7 million, of which 47% are based in agriculture (CIA, 2017). Unlike Bangladesh, the sheer size and diversity of migrants has resulted in a lack of official statistics on Indian diaspora in general and regarding temporary migrants in particular. Only a few states like Kerala and Tamil Nadu have government departments that take periodic census of its low-skilled emigrants (e.g., Zachariah & Rajan, 2016).

Low domestic employment opportunities and low growth in rural economic sectors have triggered large-scale migration. Economically distressed families often opt for internal (domestic) labor migration seeking out seasonal jobs in more prosperous regions or to work in low-skilled sectors in the urban centers (Coffey, Papp, & Spears, 2015). Cross-border labor migration is the other option for families that can find funds for the member's trip abroad. Since international migration to developed countries like Singapore often requires more economic resources, it is often the better off workers who can tap into these opportunities.

Transoceanic Indian indentured labor migration has been well documented from a historical perspective, both within and outside the British colonial frontiers (Atabaki, 2015; Emmer, 1986). Large scale labor migration from India to Southeast Asia started with the advent of British colonialism. Under the frameworks of indentured and *kangani* or contractor-oriented recruitment, early migrants from the southern coast of India were deployed to build infrastructure and to work on plantations in the British Malaya.

Moving into more recent history, Indians have migrated to the Middle East (i.e., the Gulf Cooperation Council [GCC] countries of Bahrain, Kuwait,

Oman, Qatar, Saudi Arabia, and UAE) since the oil boom in early 1970s. There are wide variations in the conditions provided to workers in the six countries of the GCC in terms of wages, number of working hours, and work conditions. However, there seems to be consensus among scholars that the regulatory systems are more abusive towards low wage workers in this region (Gamburd, 2009; Kathiravelu, 2016). Restrictive laws on naturalization and "tyranny of citizenship" (Kathiravelu, 2016, p. 59) translate to guest workers being exploited by the state, and often abused by individual employers.

As noted in the case of Bangladesh, cost of migration still remains high for low-skilled Indian workers, owing mainly to the presence of unregulated intermediaries. A study conducted by Breeding (2016) among recruitment agencies found that there was a large-scale discrepancy between actual Indian government policies and their implementation related to recruitment procedures. As per government guidelines, a recruiter can charge only between US$150 and 300 from a job candidate. Whereas, on the ground, the potential emigrant pays as much as US$800 to 1,000, an amount that can be equivalent to one year's salary for many. Breeding's study also reported that most recruiting agents "fished candidates" in remote villages, without disclosing complete details of the jobs. Exploiting the loopholes in the recruitment policy, the agents provided incorrect or incomplete information on workers' rights, their working and living conditions abroad, and the issues associated with the return migration. Breeding (2016) observes that the "idea of empowering job candidates about their rights and providing full information is non-existent" (p. 197).

While abusive regulatory regimes in the receiving countries and unregulated recruiting methods in the sending country directly affect emigrants, local sending regions also face some unintended negative consequences. For example, it is established knowledge that transnational migrations have played a major role in transforming the quality of life of people in migrant-sending regions in India. Remittances from low-skilled migrants contribute substantially to economic and socio-cultural development (Sahoo, 2016). However, studies have also noted that the entrenched forms of inequality in India fail to narrow down even when a region attracts large-scale remittance. This is evident in the case of increasing consumerism in states like Punjab and Kerala. Those who are outside the remittance net are increasingly affected by the economy and the markets adjusted to the remittance phenomenon. This is evident in the case of increased wages for low-skilled labor in migrant-sending states like Kerala. The low-skilled labor emigrated abroad is increasingly "replaced" by low-skilled workers from nearby states. Pointing to the problems of the federal system, Zachariah and Rajan (2009) argue that the benefits from emigration to the state population is, thus, difficult to establish. "The so-called spin-off effects benefited outside workers more than

Kerala workers. Thus, emigration, which initially helped unemployment in Kerala, has in fact, become a significant factor for the higher unemployment situation in the State at present" (Zachariah & Rajan, 2009, p. 269).

In sum, the pull exerted by wages in capital-rich countries in the Middle East and Southeast Asia have increased the rate of Indian low-skilled emigration in the recent decades. There have been occasional slowdowns in the migration rates, for example, at the time of global economic recession in 2008. Despite this, the cross-border movements have steadily increased. The amount paid to private employment agencies still constitutes a major portion of the migration cost in the low-income categories. Recruitment agencies have increasingly found loopholes in the Indian regulatory system such that they withhold information from job candidates and eventually trap them in a web of debts and liabilities. As regards the benefits from migration, a complex picture of widening inequality in the midst of increased consumerism and decreasing availability of low-skilled labor in sending regions is reported in literature. This provides a setting to explore the broader question of how ICT are helping reduce the cost of migration at different levels.

DO ICT LOWER THE COST OF MIGRATION?

We noted in the last section that low-skilled cross-border migrants consider migration as a step towards boosting their income and as a way to hedge against potential income loss for the family in the future. Based on neoclassical economics, low value for labor with regard to capital results in lower wages, whereas high value for labor with regard to capital results in higher wages. This wage differential determines the movement of labor force from capital-poor to capital-rich countries. The discussion above showed that on the whole, this remained the trend for most low-skilled migrant workers in Bangladesh and India crossing borders to capital-rich countries.

Individuals make rough cost-benefit calculations considering the higher wages that they can earn in the new country, given their skills and the cost associated with migration. This cost-benefit analysis might factor in both direct and indirect costs of migration. Direct costs of migration involve the immediate financial costs such as the intermediary fee, transportation costs, and the settling in expenses in the host country. Indirect and long-term costs are often unaccounted for but include the intangible effects on family and loved ones left behind. Scholars use the concept of *care drain*, among others, to understand the consequences of absent parents on their children (Barrett & Mosca, 2013; Ntale & Duma, 2011). Indirect costs also include the psychological costs involved in cutting old ties and forging new ones in the destination country. The discussion of ICTs vis-à-vis migration considers both direct and indirect costs. Indeed, the ease of ICT-supported information flow has reduced both direct and indirect migration costs, but with important exceptions that challenge any kind of euphoria related to techno-determinism. The following sections discuss these issues.

At the Individual Level

This section examines the effect of information flows at the individual level. Low-skilled migrants have gained greater access to ICT since the advent of feature phones and now smartphones, in the last few decades. Is it possible to ask how widespread is the use of this technology? How have the low-skilled migrants benefited from the use? And, does it help reduce the cost of migration, which is crucial to this segment of the population?

ICT helps migrants imagine and reach beyond the national boundaries into transnational spaces (Burrell & Anderson, 2008). Before they move into a new country, it has become a common practice for potential migrants to seek information on jobs, lifestyle, and other cultural aspects of the destination. ICTs are helpful tools to make such information available to potential migrants in the form of discussion forums, web-sites, and emails. This is more evident in the case of professionals and high-skilled expatriates. Take the example of skilled Romanian professionals studied by Nedelcu (2012). These professionals developed a "transnational habitus" as they prepared to migrate to Canada, involving sifting patiently through the vast amount of information available on the internet, and choosing the most useful information needed to facilitate a seamless transition. Informal intermediaries and touts were weeded out, as these professional migrants got access to firsthand, and perhaps first quality, information. One respondent in Nedelcu's study recalled:

> On the day I arrived in Toronto, I knew exactly what steps I had to take in the following weeks. I'd even worked out a day by day schedule to deal with administrative matters more efficiently. (p. 1348)

Studies on Zimbabwean migrants in the UK also noted how the migrants created communities online to overcome their marginalized experiences and to participate in nation building (Manase, 2013). The technology enabled them to access information on migrant experiences in the host country and news about political events unfolding back home. In addition to playing a significant role in providing opportunities for the migrants to establish networks of "economic survival," the internet helped in sending financial and social remittances—in terms of ideas and cultures—back home, helping the social and economic development of their home country (p. 42).

However, among labor migrants there have been differences in the way in which benefits are drawn from ICT use. Not all sections of migrants have benefited equally from the opportunities to find information on the internet. Female domestic workers or low skilled migrants from less-developed regions often have fewer skills and economic resources with which to exploit ICT. Xiang and Lindquist (2014) noted how Singapore-bound factory workers from China were duped by the recruitment agencies. Despite paying huge amounts in intermediary fees, these migrants were not provided accurate information about their jobs, and indeed on the challenges that awaited them, in Singapore. The information provided by the recruiter was fudged, overblown, and far from reality. In cases such as this, ICTs hardly helped individual migrants overcome barriers in information flows.

Another dimension of ICT use is communication with family members after the individuals have migrated. In many cases, ICTs facilitate this communication (Leifsen & Tymczuk, 2012; Madianou & Miller, 2011). However, in other cases, a more nuanced picture is presented. A study among undocumented Honduran migrants in the US noted that the migrants who worked in the US generally hid their stories of poverty and hardship from their families (Schmalzbauer, 2008). These migrants "relativized" their communication with family back home, choosing to hide the realities of their life in the US. These gaps in communication intensified misunderstandings and imbalances in the Honduran transnational families. The children in these migrant families aspired for and imitated middle-class lifestyle, which seemed to be the situation of their non-resident parent. This was, however, in stark contrast to the migrant parent's reality in the US. Further, the immigrant was under pressure due to the demands for material goods from his or her family. Schmalzbauer (2008) notes that "the combination of television and the transnational imaginations that are born through the flow of social and economic remittances make transnational youths uniquely vulnerable to popular culture messages" (p. 339). Thus, the information flows over ICT and via other channels have not been truly transparent and have not bridged intergenerational gaps in families.

This challenges the general notion that information flow facilitated by ICT have reduced the cost of migration for individuals and made it easier to maintain contact with families. The above examples suggest that this is not always the case.

At the Institutional Level

From a macro perspective, various institutions related to migration, from government to private agencies, have streamlined their processes using ICT. This has facilitated flow of information across government departments and transnational agencies involved in migration. This section elaborates on these changes from the perspective of cost reduction and information exchanges. It also includes critical perspectives on these changes, which argue that institutional level changes have resulted in greater racial stratifications and greater border controls.

During the last few decades, governments of sending countries have actively promoted the "export" of migrant laborers.[2] ICTs have played into these policies. While cross-border movements have reached unprecedented levels in recent decades, ICT has provided the technical basis with which governments can administer this. ICT allows them to process the applications, match the needs of employers with eventual migrants, screen people at border crossings, and control the visas of migrants. Government departments have also collaborated with private agencies to provide ICT-based training for potential migrants (Silvey, 2007).

For-profit institutions, such as recruiters and travel agencies, use ICTs to streamline their businesses and achieve greater profit. Indeed, ICT has increased the capacities of private actors in the day to day business of recruitment and transport, and banking in the case of remittances. This has resulted in a major cost reduction at the institutional level. Further, these developments have made it easier for people to migrate.

However, a critical perspective argues that the cost reduction on the institutional side has in fact curbed individual freedom, increased racial stratifications, and enhanced gender and class divide among migrants. Silvey (2007) observes that the statecraft is spatialized at a port of entry like Jakarta, Indonesia, wherein "states exert multiple overt forms of discipline and regulation" (p. 266). The low-skilled migrants returning to their home country after many years of labor overseas, thus, face an uneasy tension with the state procedures, including falsification of documents at the hands of corrupt officials.

McKeown (2004) argues that mass migration as a global phenomenon is intricately linked with the development of sophisticated systems of migration controls. This has resulted in the portrayal of migration as a "problem," especially when politicians, governments, and bureau- and technocratic circles exploit the issue for political gain. Browne (2010) calls the process of selective screening of races "digital epidermalization," wherein "identification documents, such as . . . passports, not only codify gender, race and often citizenship, but they also help to organize understandings of security, the nation and its material and discursive borders" (p. 132). These criticisms notwithstanding, there are attempts to increase interoperability across borders and sharing of biometrics database between different states (Lyon, 2008).

While ICTs may rationalize the migration process for governments and legitimate businesses, they also facilitate the work of less authentic commercial actors. The recruitment of migrants is often mired in informal businesses involving quasi-legal brokers and touts. The countries who send migrant workers often lag behind in their ability to check the operations of informal brokerages. In some cases these brokerage firms do not clearly differentiate between willing consent for migration and forced migration. ICTs have been a boon as well as a bane in this respect (Xiang & Lindquist, 2014). Studies show that in some cases, with the aid of ICTs to, for example forge documents, agents become illegal human traffickers. Kanchana (2016) reports the existence of parallel informal structures in South India, where recruiters in collusion with emigration officials send out female domestic migrants on forged documents. Breeding (2016) reports that despite advancements like the online registration system introduced by Ministry of Overseas Indian Affairs, the illegal activities of recruiters remain unchecked. Recruiters misrepresent the potentials of the work to a prospective migrant, discounting the dangers and overselling the eventual wages (as we noted in the case of recruiters "fishing" for job candidates in India). On the positive side, ICTs facilitate access to greater information flows that can be used to counter these predatory practices.

In general, the widespread digitization of the migration apparatus at the governmental and the commercial levels has resulted in better control of cross-border movements, enhanced profits for recruiters and travel agents, and enhanced ability to move remittances back home. At the same time, critical theorists have noted the "racialization" of the immigrant/minority groups and certain chances of corruption within government machinery. On the whole, ICTs have facilitated the functioning of the migration apparatus.

DIGITAL DIVIDE

The final issue in this chapter is the migrant's differential access to communication technologies, as well as their differential skill with which to use them. The concept of divide is used to explain differential access between different segments of the population in general—the poor and the rich, young and old, male and female, minority and majority communities, and so forth. In the international migration context the concept can be used to explain differences between different classes of migrants from the same origin, as well as migrants originating from different countries. In other words, digital divide can be examined at the inter- as well as the intra-national level. In the international migration context, the effect of nationality on digital divide is often mediated by factors like education level, income, and race (Benitez, 2006; Sanchez & Salazar, 2012). Among migrants from the same sending country, there are digital divides based on economic resources and skills (Baron et al., 2014).

Digital divide curtails migrants' opportunities for job mobility, access to information and social networks, hampering their overall well-being. Cartier and others (2006) identify an interconnection between scarce economic resources of migrants and the emergence of low-cost ICT like feature phones and prepaid mobile phone cards. The concept of "have-less," thus, cuts both ways for ICT and for migrants in a "dialectical" manner, as it captures relations between "practices of have-less users and the evolution of have-less ICTs" (Cartier et al., 2005, p. 11). The level of analysis here is a population in general, with greater emphasis on changing mobile phone cost and service cost from a macro perspective. However, most studies are restricted to understanding the divide at the individual migrant level.

Investigations into the information behavior of Filipina domestic workers in Singapore showed that a large number of them did not possess skills with which to access information on the internet (Sibal & Foo, 2016). The divide was noted even among workers with higher educational status. Despite their qualifications, they lacked the skills to use basic e-services such as passport renewal, document processing, and online queuing. This is exacerbated by the fact that government websites have poor security features and rarely have updated information. This results in low-skilled migrants preferring informal sources of communication over official channels (Essam, 2015).

Usual demographic variables that reflect social inequality such as race, gender, education, and employment status are also valid in the case of digital divide. Among Hispanic communities in the United States, English proficiency co-varied with the adoption and use of computers and the internet (Ono & Zavodny, 2008). In fact, both natives and immigrants living in non-English speaking areas were less likely to use ICT in their day-to-day lives than those living in English-speaking areas. A study conducted in Australia in 2007 showed that immigrants who are more proficient in English are more likely to use a computer (Chiswick & Miller, 2007). Similarly, Baron and others (2014) showed that English language proficiency is as essential as ICT skills for undocumented Hispanic laborers. They note that to survive in a city like Seattle: "Computers, phones, transportation, and English are essential for the daily survival of immigrant day laborers to adjust to the uncertain and often challenging situations they face in this city" (p. 99). Thus, in order to gain a nuanced picture of life of immigrants in the new country, it is important to understand the use or non-use of ICTs based on immigrant demographics.

That said, scholars observe that even low-income immigrants in a developed region are better off in terms of gaining benefits from ICT use, compared to their counterparts in the less developed native countries (Sibal & Foo, 2016). In such cases, migration itself is an empowering experience for these workers at the lower economic stratum, despite their many deprivations.

SUMMARY

Compared to the labor migration frameworks of the nineteenth and twentieth centuries, ranging from indentured labor to contractor-controlled recruitment methods, today's migrant regulatory systems are considered more humane and egalitarian. However, grounded realities reveal the existence of hierarchically structured mechanisms that disfavor those at the low-skilled categories. The rules in many countries follow a gendered framework, discriminating against the female domestic workers and providing the least safeguards to low-skilled blue-collar migrants.

Turning to ICTs, we considered their use within the larger theoretical framework of cost-benefit analysis done by individuals and families as they decide to move/or send a member to a foreign country for work. At the individual level, an unprecedented number of people have adopted ICTs. This is even the case among low-skilled migrants, a class that previously has not had access to these technologies. In stark contrast to the past, this class has been able to connect itself with the outside world through a range of ICTs. This has resulted in positive outcomes, such as increased access to information on job opportunities, legal instruments, healthcare, and social support from co-ethnic social networks and the ability to remit money back home. Our review also revealed that ICT facilitated information flows in the migration apparatus by incorporating new and advanced methods of screening mechanisms and effective border controls. Examining the theme of digital divide, we noted that some segments of low-skilled migrants have gained little from ICT due to economic or skill deficiencies. The next chapter examines intercultural interaction mediated by mobile phones.

NOTES

1. According to a recent World Bank report, remittance to India was US$62.7 billion in 2016 (2.8% of GDP), while it stood at US$13.7 billion for Bangladesh (6% of GDP) for the same year (World Bank, 2017).

2. Silvey (2007) notes that the Indonesian government sought to send 1,250,000 workers overseas every year during the period 1994 and 1999. This increased to yearly 2,350,000 workers during the period 1999–2004. Similarly, as of 2016, 2.2 million Filipinos are working in foreign countries at least for six months (Overseas Filipinos, 2017).

Chapter Three

Social Science of Technology Appropriation and Impact

Studies on the interaction between society and technology have followed diverse approaches, ranging from technical deterministic to domestication.[1] There have been extensive debates on the utility, relevance, and explanatory power of these approaches. This chapter briefly touches upon these debates and contextualizes them by examining the literature on mobile phone use among migrants. Based on this, the circular model of mobile phone appropriation is adopted to guide the research.

MIGRANTS AND MOBILE COMMUNICATION

Major approaches followed by mobile communication studies include adoption, impact, social consequences, appropriation, and domestication (Donner, 2008; Katz, 2006). There are overlaps in the conceptual terms used by these approaches, as they explain phenomena with similar import. For this reason, no approach can claim exclusivity. Each approach attempts to bring in more sophistication and subtlety in analysis. This is particularly needed in the case of mobile phone, which embeds newer functionalities at a rapid pace.

Adoption Studies

Adoption is the most rudimentary matrix with which to understand mobile phone diffusion among migrants. This follows a purely quantitative and straightforward method of assessing the rate of diffusion of ICTs among migrants. Vertovec (2004) assesses this diffusion by the rate of change in the quantity of mobile phone calls channeled from and to migrants. The purpose is to find a correlation between ownership of mobile phones and migrants' social mobility. This approach has a wide following when used in studies among the general population. Models like Technology Acceptance Model help in understanding the attitudinal, economic, social, and technology-related factors that determine adoption. Cross-cultural studies on ICT adoption also focus on the binary question of "ownership and non-ownership," in order to assess a country's relative position in the ICT-adoption scale.

The adoption approach helps provide a baseline upon which to build technology studies, but it fails to capture the nuances of mobile phone use. Adoption studies are not, for example, concerned with the sustained use of a technology, or the phenomenon of users dropping out after initial adoption. This failure of adoption studies was most notable in the field of *ICT for Development*, where it became known that many development projects failed to deliver after initial takeoff. Due to economic and social constraints, most users in the developing regions, trained by aid workers to use ICT in their day-to-day life, stopped using it when the trainers left. Conversely, this approach also fails to explain why a low-income population spends unreasonably high amounts on mobile phones, even by cutting down on other essential expenditures.[2] In the migration context, studies have noted that low-income migrants continue using mobile phones, spending a good percentage of their monthly income on it, but without gaining monetarily from this communication. In short, adoption approach hardly looks beyond the access/no-access or ownership/non-ownership question, nor does it explain the qualitative use of a technology.

Impact

The second approach—the impact approach—follows a functional route, as it mainly focuses on the benefits to users such as economic empowerment and opportunity enhancement. Here, the unit of analysis is generally individuals and groups. There has been increased focus on the developmental outcomes among migrants as a result of mobile phone use. Studying the migratory movements from China's inland provinces to its industrial zones in the South, Jack Qiu (2009) observes that the informational needs of a burgeoning section of migrants has fueled the growth of working class ICTs. Even while leading a precarious economic life, the workers nevertheless spend money on mobile phones and calling cards, which they consider as a means to enhance opportunities in the competitive urban milieu:

> Each of these many phone numbers [on the walls in Guangzhou or the back streets of Shanghai, advertising various services] represents a have-less individual or a group of them who do not have a stable job but can afford a mobile phone. (Qiu, 2009, p. 108)

Similar studies from across the developing regions provide a rich description of the benefits gained by migrants as a result of mobile phone use.

Historically, the adoption and impact approaches were widely followed by studies on media and diaspora. Ownership of traditional ethnic media like print, radio, and television, and the consumption of media content by ethnic immigrants were topics of scholarly interest (Fathi, 1973; Hwang & He, 1999). Ethnic media were considered to be helping in cultural transmission, in addition to acting as a community booster and a sentinel (Viswanath & Arora, 2000). Ethnic media was the most natural ally for immigrants in their adaptation and social mobility in the host country. On the political front, much like the social media-triggered diaspora demonstrations in the contemporary world,[3] yesteryears' ethnic media galvanized the diaspora towards protests and mass movements. In short, there has been sustained interest on the impact of media and communication technologies on diaspora's social, political, and economic spheres of life. However, the impact approach is narrowly focused on identifying the direct consequences of technology use. It is rather blind to the deeper political and social structures that determine the production and consumption of the technologies.

Social Consequences

The social consequences approach—the third approach—has a broad mandate. It seeks to understand the effect of mobile phone on larger social structures. This is in line with studies that, for example, looked at the long-term societal impact of introducing the steel ax among Australian aboriginals (Sharp, 1952), and the cultural changes brought about by television (Salomon, 1985). Debates on the politics and ideology of designers embedded in the technological artifact are common among studies following this approach. Works of Lewis Mumford (1934), Langdon Winner (1980), Ruth Cowan (1976), and Raymond Williams (1974) deal with technologies from this perspective. Similarly, policies related to curbing the negative impact of technology are framed at this level.[4]

More nuanced readings, however, call for identifying the reciprocal relationships between users and the technology. The question "how technologies trigger social changes" is being increasingly balanced with an equally pertinent question "how society determines the character and use of a technology." This is true in the case of today's advanced ICTs like mobile phones, which can be ascribed new meanings depending on the socio-cultural background of the users. In fact, it is widely acknowledged by theorists of social constructivism that technologies are not merely artifacts that follow an inevitable path of diffusion in society. On the other hand, technology is a text that is open to interpretations by its users. Today's smartphones and internet have multiple affordances, which allow for a wide range of interpretations. A combination of interpersonal channels and converged multimedia on these devices allow for mediated, dyadic, and group communication. This implies that the mobile phone facilitates social interaction in a way that other "stand-alone" technologies such as newspapers and TV do not (Ling, 2008, 2012). Hence, an approach that takes into consideration the multi-layered nature of the ICT platforms, and the reciprocal relationship between users and technology, is needed to guide a more sophisticated analysis.

Appropriation

The appropriation approach—the fourth approach—helps make a multi-layered analysis as required in the case of ICTs. Appropriation approach focuses on functional (instrumental) uses of ICTs, the cultural determinants of use, and the social and cultural meanings ascribed to the uses. This approach draws more from cultural studies and emphasizes that the meanings ascribed to the technology are as important as its functional uses. Take, for example, a migrant worker in our study who uses two mobile phones—one a feature phone with a 2G connection[5] and the other a smartphone with more advanced features. The feature phone is a function-oriented device that is meant to be used in the harsh conditions of the construction worksites. The smartphone, on the other hand, has greater symbolic value, and the functions assigned to it are more sentimental in nature—to be kept safe in the room, to be used for listening to songs after a day's hard work, and more importantly, to talk only with family and close relatives. Appropriation approach helps address these complexities involved in meaning making and user-technology interaction.

Appropriation is the phenomenon that takes place after adoption. It involves not only redefining the text that is inscribed in the technology by the designers, but also a kind of subversion of the technology to meet the context specific demands of the users. In mobile communication studies, a notable example is that of users in the favelas (slums) of Rio de Janeiro who alter the technical features of the mobile phones in order to evade attention by authorities (de Souza e Silva et al., 2011). In sum, appropriation provides a key to understanding (i) the way in which users read technology as a text; (ii) the functional or instrumental uses of the technology; and (iii) the symbolic meanings derived from the possession and use of a technology.

Domestication

A fifth approach—the domestication approach—needs mention here, as it overlaps considerably with the appropriation approach. Domestication focuses on the incorporation of technology in the socioeconomic sphere of the household. It focuses on the resulting changes in the social relations within the household. It understands ICT adoption in the context of social conditions of the individual (Silverstone, Hirsch, & Morley, 1992). The negotiations and interactions associated with the technology's acquisition and its ongoing role in the home or the social group have been the focus of this approach. It thus combines technical determinism and social constructivism. In addition, this approach takes into consideration the role of individuals who influence the user's interaction with technology—for example, the case of mobile phone use of teens controlled by parents.

Domestication approach identifies several steps in the adoption cycle. They are *imagination, appropriation, objectification, incorporation,* and *conversion. Imagination* is the stage at which a technological artifact or solution enters the potential user's mind (Silverstone et al., 1992). *Appropriation* is when an object leaves the commercial world and enters the sphere of life of its user. *Objectification* is when users identify themselves with the technology and decide the various modes of consuming the technology. *Incorporation* is when the technology is assigned a particular role in an array of objects. Incorporation also implies the "temporal assimilation of the objects into time structures and routines" (Ling, 2004, p. 29). *Conversion* is when the social effect of the technology becomes apparent, as also when the user becomes aware of this effect.

As the reader would have noticed, there are considerable overlaps in the way these approaches explain the phenomena related to technology adoption and use. We prefer to follow the term *appropriation* for the following reasons. First, the study included here is not a longitudinal one, so the stages leading to mobile phone adoption as well as those following it are not recorded in a chronological manner. Second, we were mostly interested in the current mode of technology use, in order to understand how this use interacted with the different modes of migrant adaptations. Thus, it is mostly a snapshot of mobile phone use at any given point in time. Third, compared to terms like domestication, which has the undertone of family and household, appropriation explains the life situation of low-skilled migrants who live singly in shared rooms and dormitories.

We adopted the circular model of mobile phone appropriation ([Karnowski, von Pape, & Wirth, 2008]; a simplified version of Wirth and others [2008]; hereafter "the circular model"; see figure 1 in appendix B) to guide our research. The circular model integrates a range of approaches in diffusion and adoption research like frame analysis, cultural studies, and uses and gratifications. As a result, issues like the symbolic appeal of new communication devices for its users and meta-communication related to the process of appropriation have received increased attention in this model. In the following sections, we introduce the model, its different features, and the relation between different components in the model. We review studies in mobile communication that have followed different approaches and examine how they support different features and relations hypothesized in the model.

MOBILE PHONE APPROPRIATION MODEL

Three components of the model—*usage and handling, prestige and social identity,* and *meta-communication*—explain the process involved in appropriation. The first two can be broadly defined as the functional use of the phone and the symbolic meanings generated by its use. Meta-communication is communication on communication, which involves one's own as well as others' perception of the social effect of mobile phone use. This is akin to *conversion* in the domestication approach, which is a phase at which the user is aware of the effects of technology use and begins to communicate them with others. We turn to these components more elaborately in the following sections.

Functional Use and Benefits from Mobile Communication for Migrants

The component of the circular model that focuses on the functional (instrumental) use of the mobile phone is *usage and handling,* denoting the ways of using the mobile phone for practical purposes. It also denotes the ways in which mobile phones are melded into users' lives. Functional uses are determined by factors like technology skills, communication needs, and financial resources of the users. We explain them in detail below.

Technology Skills

To operate any technology a set of skills is required. Mobile phones and ICTs are no different. Lack of technology skills on the part of certain segments of the population is a theme that has received wide attention in ICT literature. Older generation, people with low education, women, and low-income segments of the population are the most vulnerable. Among migrants, professional expatriates and high profile immigrant entrepreneurs are proficient with most of the functionalities of the mobile phone, whereas low-skilled migrants are less proficient. Exploring how ICTs helped professional expatriates develop a "transnational habitus" even prior to migration, Nedelcu (2012) sketched the experience of a number of Romanians in Canada. Prior to their journey, these expatriates had regularly searched the internet, sifting patiently through the vast amount of information before finally zooming in on the most useful information they needed in the migration process. One respondent in Nedelcu's study recalled:

> While I was still at home, I tried to get in touch with people in Canada. First I found a discussion group for Romanians in Vancouver . . . I found guidelines for newcomers and a section dealing with practicalities: preparing for departure, what to bring in your luggage, what to do during the first week, etc. (p. 1348)

Discussion forums on the internet turned out to be a vast source of information for these professional expatriates. Whereas, investigations into the information behavior of Filipina domestic workers in Singapore showed that a large number of them did not possess the skills with which to access information on the internet (Sibal & Foo, 2016). The divide was experienced even by workers with higher educational status, who lacked the skills to use basic e-services.

Communication Needs

Migrants have diverse socioeconomic backgrounds and job profiles. Hence, their communication needs also vary. Most common among the communication needs is the regular contact migrants maintain with their home country, as the vignette of Francis we noted in the opening paragraph of the book. Such contact includes accessing news on political and social affairs as well as news of personal interest from home countries. Vertovec (2004) described the cross-border communication of migrants using mobile phones as "discussions across a kitchen table . . . in real time across oceans" (p. 222).

Communication needs at the workplace also lead to migrants' reliance on ICTs in general and mobile phones in particular. For instance, early studies showed that male low-skilled migrant workers in Singapore watched technical videos while at work when they had to fix problems at worksites like a welding crack, or a leaking pipe (Chib & Aricat, 2012). Mobile telephony also helped in greater coordination within a team of low-skilled workers, which has resulted in easy completion of tasks, and higher amount of leisure for the migrants. On one hand, such instrumental uses have the potential to increase work productivity, whereas on the other, the time gained as a result of better coordination is used for leisure. Similarly, task equivocality—personal focus and possibility of instant feedback (Lin, 2003)—and richness of the medium help produce "realism" in mobile communication.

By helping fulfill the communication needs, the mobile phone empowers migrants who are vulnerable to workplace harassment and maltreatment. Law and Peng (2008), in their study among in-country migrant workers in China's

Guangdong province, narrated stories of low-skilled workers in a historical perspective. In the pre-mobile era, the workers were not often in contact with their kinsmen. In addition, they lacked knowledge of employment opportunities elsewhere in the province. Owing to this disconnection, the employers exploited, mistreated, and even assaulted them. But, as the workers began using mobile phones, they found a plethora of job opportunities. According to Law and Peng (2008), they were no longer afraid of being fired since they knew that they would be able to find work elsewhere in the same province thanks to their well-connected networks.

For foreign domestic workers in Singapore (most often women), the most pressing communication need is to evade the control of their employers and get connected with their co-ethnic friends' network (Lin & Sun, 2011). Short messages, often sent furtively, allowed for such connectivity away from the surveillance of the employers. Thomas and Lim (2011), in their ethnographic study among Filipino and Indian domestic workers in Singapore, observed that mobile phones brought relief to the domestic workers from their repetitive daily chores and drudgery. Calling and sending text messages to their family members back home became private rituals via the phone. This ritualistic communication also helped them, at least temporarily, escape the physical constraints they faced in their life as domestic workers within the four walls of the employer's house. Mobile phones allowed the domestic workers to develop intimate relationships, although often supplemented with intermediaries and face-to-face interactions:

> The availability of mobile phones helps the foreign domestic workers improve their isolated working conditions and living status, as well as provide a communication channel to maintain close ties with their families overseas and broaden their local social networks. (Lin & Sun, 2011, p. 191)

Mobile phones empowered migrant workers psychologically and practically in their interactions with authoritative figures.

Migrant mothers have the communication needs based on the demands of mothering their own children in distant home country. In this case, regular communication over internet and mobile phone can only partially fulfill the mothering demands (Wilding, 2006). Emigrant parents from Jamaica in the United States and Britain "participate[d] in the day-to-day affairs of their children" with the help of mobile phones (Horst, 2006, p. 149). However, the separation between migrant mothers and children admittedly incurs high indirect cost to families involved.

Thompson (2009) identifies the ironic situation of mobile technology's rapid diffusion into arguably the most immobile population in Singapore, the migrant workers. In contrast to foreign talent who are mostly globe-trotters, the workers have limited mobility in their professional life. Despite this, the workers were one of the early adopters of the mobile phone:

> As utilitarian devices, handphones are transformed by the uses to which foreign work put them—for example, from devices for mobile people to devices that help people overcome constraints of their own restricted mobility. (2009, p. 360)

While Thompson's observation points at the differences in the communication needs of migrants, "restricted mobility" may not be generalized to all low-skilled migrants. A section of migrant workers in this study moves regularly from one construction site to the other as part of their work.

Mobile phones have also become effective communication channels through which migrants can protest against low wages and poor working conditions (Ngan & Ma, 2008). As conduits of latest data and information, mobile phones helped workers compare their work conditions and remuneration with that of their peers in other regions. If inequities were found, protestations were often aired through mobile phones as short messages and snippets:

> The real-time conversations and virtual face-to-face interactions available through this technology have strengthened their sense of community and reinforced their local identity, which would otherwise be threatened in unfamiliar host cities. (Law & Chu, 2008, p. 44)

That said, mobile phone use has not always resulted in benefits for migrant users. Wallis (2008) looked at the ways in which female rural-to-urban migrants in China negotiated their identity and agency in the urban environment using new communication technologies, especially internet and mobile phones. Owing to limited literacy, financial resources, technical expertise, and their gender, the migrant women could be seen as being on the losing side of the digital divide. While documenting the ways in which these young women tried to overcome the structured patterns of social hierarchy and power, Wallis nevertheless argued that ICTs "at the same time can potentially preserve, or even strengthen, these same [patriarchal] structures" (2008, p. 200). Wallis observed that like other technologies, mobile phones also reinforced existing asymmetrical power relations and created new modes of exploitation and control.

Among migrant mothers, mobile phones created dialectical tension between different identities of transnational mothering, which was resolved by the latter with the help of unique and counterintuitive communication strategies like actively disengaging from mobile phone use (Chib et al., 2014). Vancea and Olivera (2013) note the emergence of an e-migrant wherein technological and geographical spaces are constantly reconfigured by new media technologies. However, for migrant mothers in their study, the social spaces by far remained the same. In the estimation of Vancea and Olivera (2013):

> E-migrant women cannot always escape family and community-based social networks, and continue to maintain their traditional roles as mothers, daughters, sisters, and spouses from a distance. (p. 199)

In sum, literature widely supports the empowering potential of mobile phone among migrants. Based on job profiles, gender, and socioeconomic status, the communication needs of migrants vary. This section summarized the communication needs of migrants in a range of job profiles and in a wide range of geographical locations. Studies have also looked at the ways in which mobile phones have produced additional layers of power negotiations in migrants' day-to-day lives. Low-skilled migrants also face limited financial resources, which has a negative impact on their mobile phone use. We discuss this topic in the following section.

Financial Resources

Studies have widely discussed the economic resources of low-skilled migrants, which have a bearing on their spending on ICTs and the benefits they gain from it. Evidences suggest that the cost of voice calls across international borders has reduced tremendously in the last one-and-a-half decades. In addition, communication is now channeled via internet-based voice platforms like Skype and Google Talk. In a study conducted by Rahman and Lian among Bangladeshi migrant workers in Singapore in 2005, 18% of the respondents made no telephone calls to their home country owing to prohibitive call charges. This situation has changed in less than a decade; there is hardly any migrant worker without a mobile phone at present in the location studied by Rahman and Lian. In addition, the advanced features of mobiles have made rapid inroads into the migrant population.

However, any attempt to draw a linear causal relationship between migrants' financial resources and their mobile phone use may be simplistic, given the complexity of ground realities. Increased transnational movements incur indirect cost to families, as is evident in the case of migrant mothers. In addition, there are workers who are affected by low wages and illegal status. In the US, but also in countries in Western Europe, the documented and undocumented migrants who arrived after the economies' shift from manufacturing to services suffered from low wages and job insecurity at the hands of sub-contractors. These workers are deterred by the high cost of transnational communication. Mahler (2001) problematizes the way in which transnational communication is "presumed and not problematized in research and publications," and calls for a nuanced understanding based on gender and power (p. 585). Schmalzbauer (2004) explored the transnational family experience of migrants from Honduras to the US and observed that transnationalism threatened family proximity, which was more evident in the case of poor families divided by migration. Tazanu (2015) examined expectations associated with remittances by non-migrants towards migrants. These were often conveyed to them over mobile phone. Tazanu noted:

> Many young people assessed the state of their relationship with migrants in terms of frequent phone calls and material and financial support they received or expected from the bushfallers.[6] (p. 25)

These examples suggest the need for a nuanced understanding of transnational communication among low-skilled migrants, beyond the frameworks of affordability and financial means.

By way of summary, this section has reviewed studies on the use of mobile phones among migrant workers, focusing on its impact on social relations and the benefits gained by the migrants as a result of the use. This functional approach has been captured by the *usage and handling* component of the circular model, suggesting that use aimed at achieving pragmatic goals—maintenance of relationships, management of daily life, distraction as a pastime—is a subset of the larger domain of appropriation. The following section elaborates on the symbolic part of the circular model.

Mobile Phone's Symbolic Meaning for Migrants

Appropriation is an interpretive schema used by studies to explain the phenomenon that occurs after technology adoption. Bar, Pisani, and Weber (2007) define appropriation as the "process through which [technology] users take something external (alien, or foreign, something given to them by others), and make it their own" (p. 3). This idea goes beyond the mere adoption of the technology in question. Rather this conceptualization includes the process of integrating the technology into users' day-to-day work and leisure. Appropriation is also a political act of re-negotiation with the power structures that determined the structural and functional aspects of a given technology when it was designed and produced. The metaphor of technology as "text," which is prevalent among social constructivists, has greater importance in appropriation studies. The "text" metaphor highlights users' multiple readings of the technology and their unique ways of creating meanings in everyday practices (Grint & Woolgar, 1997). Appropriation research has focused on different populations, including older generation (Karnowski, von Pape, & Wirth, 2008); youth (Ling & Yttri, 2002); and migrants (Ngan & Ma, 2008).

The circular model postulates that the multiplicity of meanings produced as a result of mobile use, and the meanings that determine the self/social identity of the user are as important as the goal-oriented, pragmatic conceptions related to the phone. The *prestige and social identity* component highlights symbolic appropriation of the technology, "how does the user present him/herself with the innovation? Does he/she use it as a *prestige* symbol to enhance his/her *social identity*?" (Karnowski et al., 2008, p. 184). This is conceptually similar to *conversion* discussed in the case of domestication earlier (Silverstone, Hirsch, & Morley, 1992). Indeed, the mobile phone has emerged as a prestige symbol for youth (Ling, 2007), and as a marker of social identity for the mainstream society (Lievrouw & Livingstone, 2002). Mobile phone use has influenced the personal and social identities of migrants, improving their self-esteem, ensuring membership in ethnic groups, thereby helping them gain peer recognition (Chib & Aricat, 2012; Wallis, 2008).

Often, the symbolic role of mobile phones overlaps with its functional uses. Yang (2008) observed that migrant workers in the service industry in South China saw mobile phones as a tool of empowerment, as much by the technology's symbolic value as by its utility. This empowering potential of the mobile phone, however, was only one part of their overall experience of migration, which they considered as empowering in its totality. Yang observes:

> Mobile phone usage cannot change migrants' overall disadvantageous socio-economic position, mobile phones are often a tool for alleviating boredom and evading and mitigating the intrusive and overbearing demands of work supervisors. (p. 68)

We present several examples later in the book wherein migrant workers' self-esteem was enhanced when their technical prowess with the mobile phone was recognized among their peers.

In sum, the symbolic meanings ascribed to the mobile phone are often as important in the appropriation process as their functional uses. Especially in the case of low-skilled migrants, symbolic empowerment helps them endure hardships like lack of equal treatment in the host civil society and prolonged loneliness and alienation. Both the functional uses and the symbolic meanings are linked by meta-communication, which is discussed below.

Meta-communication

Linking the two phenomena of "functionally beneficial uses" and "symbolic meanings" is meta-communication, which is defined as communication on communication. Users, in the process of using a technology, reflect on their experiences, share them with others, while seeking information and advice from different sources. Among migrants, meta-communication has greater relevance, as they need to familiarize themselves with the new ICT eco-system they have migrated to. Meta-communication includes advertisements and news related to mobile services, but also interpersonal exchanges between users and non-users related to mobile devices and services.

Meta-communication is the process through which both *usage and handling* and *prestige and social identity* are interlinked, thus reflecting the circular nature of the model. This circular process is not always based on the synergy between functional uses and symbolic meanings. On the other hand, there are tensions between these two aspects. For example, respondents in our study complained about people increasingly becoming addicted to the smartphones such that it became difficult even to get an eye contact, not to speak of the possibility of initiating a conversation, with fellow passengers in a public transport. Here we find an instance of interpretations conflicting with each other—while the actual users might see the public use of the mobile phone as a kind of empowerment, the onlookers interpret it in a diametrically opposite manner. Thus, the symbolic meanings and functional uses conflict with each other, depending on who interprets them.

To summarize, different approaches help examine the diffusion, use, and effect of mobile phone in society. The approaches range from those merely making a quantitative assessment of the diffusion of ICT in society to those taking a critical approach towards the power negotiations and collective meaning making involved in the use of the technology. In the approach we follow, there are tensions and synergies between functional use of the phone and the symbolic meanings derived from the technology. This is more akin to the position of social constructivism wherein technology is a text that needs to be interpreted by users.

The review thus far has been on the interaction of mobile communication with migrant life in general, without emphasizing on the cultural implications of such interaction. We now turn to two major cultural engagements of migrants which involve communication—(i) migrants' sustained relationship with their culture of origin; and (ii) migrants' engagement with other cultures in the host society. We discuss how these interactions are played out in the mediated space of mobile telephony.

CULTURAL SALIENCE OF MOBILE PHONE COMMUNICATION

Cultural salience denotes the various ways in which a message, or the activity of communicating itself, utilizes shared symbols, language, and idiom of a particular culture. This has been a major topic of concern in cross-cultural studies, as an individual's adaptation to a new culture is determined by the amount of communication channeled from or to the new culture. This section reviews studies that have focused on migrants' communication links with home and host cultures. By mapping these communication links, we move closer to identifying the theoretical framework that has best captured these interactions among migrants, which ultimately determines their adaptation.

Prior to ICTs, ethnic media were a platform on which communication from home and host cultures converged. Studies investigating the cultural implications of migrants' use of ethnic mass media identified a tension between migrants' desire to retain ethnic identity and their need to relate to the dominant cultures of the host society. Subervi-Velez (1986) noted that ethnic media promoted assimilation of migrants to the dominant host culture only in selected socio-political domains, but otherwise propagated cultural pluralism and maintenance of ethnic culture. Zubrzycki (1958) observed that the foreign language press respected migrants' feelings of affinity towards their home culture, but also endeavored "to inculcate an understanding and respect for the institutions of the country of settlement and to explain to its readers the significance of native customs" (p. 79). In the mediated and networked environment of the mobile phone, users have greater control on the cultural content they want to subscribe to or ignore. The following sections discuss these possibilities in detail.

Mobile Communication with Home Culture

Two areas are identified where migrants' communication from/to home culture is found to be at the highest: (i) migrants' interaction with co-ethnics in the host country, and (ii) their interaction with family and friends back in the home country and beyond. We are concerned with both the one-on-one communication via mobile phone calls and text messages, and the multimedia content like audio-visual material available on smartphones. When the fieldwork for this study was conducted, migrant workers relied mostly on telephonic networks supported by 2G technology. A large part of one-on-one communication was channeled via these networks at that time. Today, one-on-one communication among migrants increasingly utilizes internet-based platforms like Skype and WhatsApp. A small percentage of migrants in our study used 3G smartphones, which helped them access multimedia content. Irrespective of the ICT platforms used, a large part of migrants' communication made their culture of origin salient. For example, a low-skilled migrant might be talking to his family in his country of origin or watching a movie in his vernacular mother tongue. The first involves one-on-one communication, whereas the second is consumption of multimedia content. Despite these differences, the language spoken, the idioms used, the events, and personalities that are referred to and the cultural symbols evoked are similar in these two cases. It makes the home culture salient. In this section, we argue that migrants' mobile communication with home culture may be explained in a pluralistic manner, shunning the uni-directional and linear approach followed by assimilation studies.

Co-Ethnics in the Host Country

Mobile phone studies among migrants have reported a range of communication ties they develop in the host country (Chib & Aricat, 2012; Law & Peng, 2008). Primary among them are the connections with members of the same ethnic community or co-ethnics, which are found to have significant effects on adaptation. Owing to migrants' lack of proficiency in host society language and their lack of information about the host country, they usually take guidance from the co-ethnics. These connections are then maintained generally throughout migrants' sojourn in the new country.

Co-ethnics belong to different formal and informal organizations in the host country, depending on their interests, demographics, and the cultural heritage to which they originally belong. Indian immigrants, for example, have associations based on linguistic identities. In Singapore, these include Bengali Association,[7] Maharashtra Mandal,[8] Malayalee Association,[9] and Telugu Samajam,[10] among others. In Singapore migrant workers come together for prayer and recreational activities organized by religious groups. Most of these are co-ethnic groupings, which allow opportunities for closer ties and development of social capital. While work and earnings take precedence in professional organizations, migrants benefit from "socializing, training, and resource-sharing" (Roldan, 2009, p. 8) by participating in informal and voluntary organizations.

As we discuss later in the book, mobile communication is central to the retention of co-ethnic networks. This can range from arranging a group event like a prayer meeting at a designated place, or two co-ethnic friends quickly catching up over a movie. The close associations between co-ethnics often extend to workplaces, wherein workers from same linguistic and ethnic backgrounds come together in a team. Research suggests that co-ethnic workers in a team have greater satisfaction in their work and have higher productivity. In the case of migrant workers, mobile communication has a major role to play for coordination within a team, especially among co-ethnic workers. This cultural salience of migrants' regular interaction with co-ethnics in the host country is one of the focus areas of this study.

To Family and Friends in Home Country and Beyond

Mobile phone has also helped migrants retain sustained ties with family and friends in their country of origin (Law & Peng, 2007; Roldan, 2009; Vertovec, 2004). Voice calls via the mobile phone, via telephonic networks, or over IP-based applications, often supported by video and short messages, are the most widely used modes of communication. The importance ascribed by migrants to regular connection with family is often rooted in their culture. Lindsley (1999) observes that Latino diaspora has a high level of communication with family since trust and family stability are two crucial aspects that mark their culture. Similarly, mutual trust and support have been identified as key in maintaining cohesion among Asian families (Yi, 2013).

Apart from the family circle, migrants' interaction with the home culture also includes other domains of life. For example, migrants participate in the political activities back home by funding political parties and utilizing their absentee voting mandate. Migrants also fund various types of developmental activities in their home countries. Migrants invest money in businesses and buy stocks in their country of origin. These transactions require constant engagement with various actors in the country of origin.

Sustained communication links have been observed between migrants belonging to the same culture—often family members or friends—but working in different countries outside their country of origin. This is theorized by the concept of transnationalism, which explains migrants' connections, cultural engagement, and business transactions beyond the binary of "country of origin" and "destination." Transnationalism finds its highest expression on ICT platforms.

The review above shows that migrants connect with their home culture in a wide variety of ways—with co-ethnics, organizations, and activities in the host country; to family and friends in home country as well as around the world. There is substantial evidence to suggest that if migrant workers can interact with friends from their own ethnic community in the new society, it alleviates many of the problems associated with their cultural adaptation (Chen, 2010; Green & Lockley, 2012; Holmes & Janson, 2008). However, studies provide mixed evidence on how this home-culture oriented communication affects migrants' social integration to the host country. One key argument is that mobile communication helps develop "particularistic social bonds" (Geser, 2006, p. 10), implying that people tend to connect exclusively with others within the same social circle. In the case of migrants, this means that there will be increased ethnic cohesiveness even to the extent of a virtual disconnect from the dominant culture of the host society.

Dahlberg (2007) argued that diaspora's engagement in computer-mediated spaces can result in non-engagement with host society. This is because debates are fragmented according to the ideological position of each virtual group. Sunstein's (2001) "deliberative enclaves" also denotes a similar phenomenon of non-engagement between people of different cultures and worldviews (p. 67). Horst (2006) notes that migrants' increased ties with the home culture has posed a challenge to their eventual settlement in the host country. She notes:

> The comfortable distance many transnational migrants may have enjoyed or even required for a successful settlement in the country of migration may dissipate with the expanding availability of telecommunications such as the mobile phone. (p. 155)

Studies have observed that cohesive ethnic groups remain aloof from the mainstream host society, thereby imposing a "ghettoization" upon themselves. Ryan and others (2008) argued that overreliance on dense networks constituted by one's own ethnic group reinforced migrants' dependency on niche markets. It worked as a disadvantage to migrants in the long run: "(F) ar from generating economic capital, tight networks of co-ethnics may signal social disadvantage and ghettoization" (Ryan et al., 2008, p. 686). Already existing ethnic networks become impenetrable for new entrants, as existing members take a critical view of their fellow ethnics newly migrating to the host country. Kathiravelu (2012) observed that co-ethnic networks can turn exploitative if family or friendship ties are involved, which makes breaking out of the network difficult. In short, tightly-knit ethnic groups may not always be an optimal condition for migrant adaptation; inter-group trust remains low and community participation is hardly attained (Portes, 1999; Portes & Sensenbrenner, 1993).

A study conducted earlier by Aricat (2015) showed that low-skilled migrant workers, due mainly to their disadvantaged position and low social standing, bonded stronger with their own ethnic communities in host country Singapore. The migrants used only one-to-one communication modes like voice calls and short messages on mobile phone with which to relate to host society members, whereas their relation to home culture was more varied. By lowering the threshold for migrant workers' mobile phone–based contact with their home culture, there was a reduced sense of belongingness to the host culture. Bereft of opportunities for long-term settlement, migrant workers rarely attempted a deeper engagement with the host society, which seriously threatened their feeling of belongingness to the host country.

We argue that migrants' interaction with the home country/culture needs to be evaluated using a pluralistic framework. This helps go beyond the linear conceptualization of assimilation. In the case of low-skilled migrants who are transient workers, the idea of assimilation may be out of place, since the policy framework does not support or foresee their long-term settlement in the host country. Owing to this reason, we have adopted a term that is more inclusive and more meaningful to the low-income segments of the migrant population, who are otherwise excluded from the discourse of integration. We use the term "acculturation" in the study to assess the strategies that the migrants adopt to engage with the host society. Communication to the host society is a key step in this process. We discuss it in the following section.

Mobile Communication with Dominant Host Cultures

In addition to the communication networks among co-ethnics, migrants also develop connections with host society members (Kim & McKay-Semmler, 2013). Historically, host society media provided a platform for migrant communities to gradually adapt to the host society. This mediation was largely achieved by helping migrants acquire host society language skills and providing them with information on education and employment opportunities. Migrants who accessed host media content were found to be better adapted to the society. However, as mentioned in the earlier section, low-skilled migrants rarely have the opportunity to associate with any of the dominant markers of the host society culture. For example, they prefer their own ethnic media over, for example, the English language press; they rarely have membership in resident associations or recreational clubs hosted by the local population. This has reflected in their lack of engagement with the host society both in offline and online spaces.

Nevertheless, there are more opportunities for low-skilled migrants to engage in multicultural teams at workplace, which facilitates their adaptation to the work organization. In Singapore, the workforce is mostly multicultural. Thus, a migrant worker from Bangladesh or India would invariably collaborate with Chinese, Malay, and Thai workers, along with native Singaporeans at the workplace (Rahman & Lian, 2005). Studies have shown that mobile communication is indispensable for coordination within work sites that are often sprawling over a considerable expanse. We draw from theoretical frameworks in the field of organizational communication to understand workers' coordination over mobile phone during work.

Studies have identified the importance of communication within organization. Irrespective of the cultural diversity of the workforce, better communication leads to good relationships among coworkers (Anderson & Martin, 1995). Employees tend to have positive attitude towards work climate and better job satisfaction, if communication between them is unhindered (Muchinsky, 1977). Mobile phone is a key tool with which to facilitate this, as workers are able to call to one another to organize deliveries or to work out the logistics of work processes. This has motivated scholars to conclude that "communicating . . . contribute(s) to employees' satisfaction at work" (Anderson & Martin, 1995, p. 251). Organizations also want to promote communication links between employees, so as to improve their commitment towards the organization, which will eventually "reduce withdrawal behaviors and turnover" (Mathieu & Zajac, 1990, p. 171).

Mobile phones have been instrumental for migrant workers to establish relationships with host mentors. According to Kathiravelu (2012), "Through their employer's connections, low-wage migrants often secure employment and promotions for family members and friends" (p. 110). NGOs and charity groups that support an informal care system for migrants seek patronage from host mentors, in order to ensure the sustainability of their work (Roberts & Portes, 2006).

There are instances that emerged in our data, as also supported by findings from other studies (Stephens & Ford, 2014), that often workers are not allowed to use mobile phones during work hours. Rather, they largely depended on the filtered information that was passed down to them via supervisors. Often, such communication reached them late and was burdened with unnecessary ambiguity. Knowledge workers, unlike manual labor, have ready access to computers and mobile devices at work and they receive crisis and emergency messages directly. This points to the issue of digital divide, albeit an enforced kind of divide, in the access to ICT at the workplace. We explore in this study how migrant workers' acculturation is influenced by their use and non-use of mobile phones to communicate with the host society and to their job. The findings suggest that adaptation to the work organization is found to be higher in the case of workers who engaged with members of other cultures in work teams. Mobile phone communication facilitated coordination among members of multicultural team. Better adaptation to work also facilitated adaptation to social life. However, there were instances when mobile phones were banned at work, mobile communication failed to solve problems in organizations, and mobile communication failed to bring workers of different cultures together. We discuss these instances in detail later in the book.

SUMMARY

This chapter reviewed five approaches in the study of mobile communication among migrants, which are rooted broadly under technical determinism and social constructivism. The appropriation approach was found to be suitable for this study owing to its ability to capture both functional uses of and symbolic meanings ascribed to the technology. A suitable model in the appropriation approach—the circular model—was chosen to guide the investigation. Three key components of the model—*usage and handling, prestige and social identity,* and *meta-communication*—were explained based on the extant literature and studies conducted among migrants.

This chapter further identified the cultural import of migrants' communication and classified it broadly as home-culture oriented communication and host-culture oriented communication. It also reviewed studies in these two areas, drawing relations with studies on ethnic and host society media. It also drew from literature on organizational communication to highlight the role of mobile communication in organization. Recapitulating the kind of connections maintained by migrants, the chapter described them as "mobile phone communication to home culture" and "mobile phone communication to dominant cultures in the host country." Despite the diversity in professional, social, and cultural persuasions observed in such connections, a pattern observed is that either they make salient migrants' affinity with home culture or they make salient migrants' attempts to relate to other dominant cultures in the host country. Theoretically, such a pattern mirrors the typology of acculturation based on a bi-dimensional model of the phenomenon. The next chapter reviews the bi-dimensional model and explains how the communication and cultural ties reviewed in this chapter can be integrated with the model to provide a theoretical basis for the investigation that follows.

NOTES

1. Ling (2004) explains these approaches and their advantages and disadvantages for studying the impact of mobile communication in "Chapter 2: Making sense of mobile telephone adoption."

2. It is observed that mobile phone helps low-income groups during times of emergencies and for even small-scale "begging" of money from their contacts. Horst and Miller (2005) write, "It hardly seems surprising . . . that the purchase of a cell phone is regarded as a priority by those who might otherwise be considered least able to afford them" (p. 762). Souter et al. (2005) observe that "the telephone is considered to have economic value by higher income groups, but to have negative economic value by lower income groups" (p. 14).

3. This needs to be qualified. In host countries that have long democratic traditions and well-entrenched diaspora—like the US and the UK—demonstrations by diaspora supporting or opposing the policy measures of home and host governments are common. However, in countries where freedom of political speech is restricted (e.g., Singapore), the chances of neither ethnic media nor today's social media galvanizing the diaspora towards a demonstration are less.

4. Incidentally, as we write this, Hanoi, Vietnam, has proposed a ban on motorcycles by 2030. A city of five million motorbikes (or, 2,500 motorbikes per kilometer)

as of 2017, Hanoi registered higher than normal levels of Particulate Matter 2.5 for 282 days in 2016 ("Hanoi plan," 2017).

5. Although 2G phones were widespread at the time of fieldwork, the service providers in Singapore have withdrawn 2G since early 2017.

6. A term used in Cameroon to denote migrants who have moved to greener pastures, even if temporarily.

7. http://www.bas.sg/

8. https://www.mmsingapore.org/

9. http://www.malayalee.org.sg/

10. http://www.sts.org.sg/

Chapter Four

Acculturation and Mobile Communication

Migration has been studied from different perspectives, among which the most relevant for this study are: first, cultural adaptation of migrants in the host country; and, second, transnational spaces created by migrants as a result of their interaction with different cultures. Although these approaches inform each other, the political, economic, cultural, and technological persuasions of each approach vary. This chapter reviews key concepts and the empirical studies that are motivated by these approaches, while identifying the relevance of technology in each case. In the first approach of cultural adaptation, individual and group characteristics are at play, as this approach considers migrants as active agents in adaptation (i.e., when they negotiate their affinities with the culture in which they are rooted [the home culture] and the dominant cultures of the host country). In the second approach, transnational spaces highlight the importance of migrants' crossing of national borders, which entails questions of citizenship, political entitlements, human rights, and migrants' role in the creation of a public sphere in the liminal space of neither here nor there.

An underlying theme that cuts across these two approaches is "acculturation strategies" or "orientations," which are defined as the disposition of individual immigrants or groups to engage with host societies. There is a long tradition of describing acculturation strategies based on immigrants' communication with host and home cultures. We draw from this tradition and discuss how mobile communication among immigrants can provide a new window with which to look at the issue of acculturation. As the chapter reviews literature on acculturation, adaptation, and transnationalism, it also draws connections with the previous chapter on migrants' appropriation of mobile phone and their communication links with home and host cultures.

Before proceeding, it is important to note that most of these concepts seemingly overlap in the way they describe similar but different phenomena. For example, acculturation and adaptation may appear the same for the public. A more technical examination reveals that acculturation is more about the immigrants' motivations and efforts towards adaptation. The latter, thus, is an outcome of the former. In this study, we propose a model with acculturation strategies and communication over mobile phone as causal factors, and migrants' level of adaptation to the host society as an outcome variable.

ACCULTURATION: TWO APPROACHES AND A TYPOLOGY

Depending on the nature of migration, the status of migrants in the new country, and the long-term ambition of the migrating group, immigrants' engagement with the host society varies. For example, those who are labor migrants, and who are sure to return to the homeland after a definite period of employment in the foreign land, tend to have a lower sense of belongingness to the host country compared to those who have greater permanence in the host society. Defined using terms such as assimilation, enculturation, and biculturalism, academic scholarship attempted to create a model for migrants' cultural experience in the host country (Gordon, 1964; LaFromboise, Coleman, & Gerton, 1993). A more nuanced approach towards the phenomenon emerged with the concept of acculturation.

Acculturation is an integrated approach for analyzing the varied problems migrants face in the new culture, and eventually categorizing and predicting the migrant adaptation outcomes (Arends-Tóth & Van De Vijver, 2003; Liu, 2007; Piontkowski et al., 2000). In the early years of acculturation studies, academic scholarship was deterministic in its approach, as it considered assimilation or immigrants' complete blending to the dominant host culture as natural and unproblematic. For example, the Chicago School in mid-twentieth century reified Americanness among immigrants. However, this trend has since been challenged by more pluralistic approaches. Currently there are two main approaches towards acculturation (Berry, 1997; Kim, 1989): an older approach of defining acculturation as a *cumulative-progressive* phenomenon and a more recent *pluralistic-typological* approach that defines acculturation in multiple and mutually exclusive categories of cultural bonding (Berry, 1980). The following sections review these approaches.

The Cumulative-Progressive Approach

The cumulative-progressive approach considers acculturation as a step-by-step process, wherein migrants adapt to the new culture with the passage of time (Kim, 1977, 1978, 1984). This approach has been characterized by a uni-directional model in which acculturation is a linear process beginning with low level of adaptation and ending in "assimilation" or one's complete blending to the new culture. Thus, Korean or Chinese immigrants entering the United States are said to be slowly assimilating to the American society, shedding their identification with their culture of origin. The uni-directional model describes acculturation in proportion to the amount of time a migrant spends in the new culture. It also follows a normative, deterministic path that maintains that a total blending to the host culture is attained only at the expense of forsaking migrants' native cultures. In this approach, since time has an additive effect on acculturation, it is called "cumulative-progressive" view.

Within this framework, there have been attempts to assess the assimilation of ethnic groups over a period of time, and to identify the factors that influence the process. Acculturation of the second-generation has been a major area of research. A dominant view of early twentieth century was that immigrants' incorporation into the professional and mainstream business sectors resulted in the upward mobility of the second-generation. Studies conducted in the 1980s and 1990s challenged this dominant view (Perlmann & Waldinger, 1997). Highlighting the role of parents' economic and educational achievements and the social capital of ethnic groups and family, Portes and Zhou (1993) showed that most minority second-generation youth in the US, instead, faced downward mobility. This "segmented assimilation" theory also highlighted factors like the long-term goal of an ethnic group in the host country and the context of social and legal reception enjoyed by the group. The cumulative-progressive approach was, thus, being reviewed based on empirical data from the labor, economic, and cultural fronts.

The segmented assimilation theory itself was later reviewed and criticized by a number of studies. In what is called the "new" assimilation theory, Alba and Nee (2003) criticized the pessimism of downward mobility posited by the segmented assimilation theory (Ali & Fokkema, 2015). They emphasized more on the institutional arrangements that determined migrant assimilation, for example, policies of the state, work organization, and the labor market in general. In one analysis, Alba and Nee (2005) explain how low-skilled labor market shrunk in the US between 1930 and 1970, as the country witnessed a surge in white-collar occupations. This structural mobility resulting from realignments in the labor sector decreased the need for unskilled labor, a sector in which immigrants were concentrated in the early part of the twentieth century. Alba and Nee (2005) observe that the ethnic youth during this period had to make a tough choice between getting enough education "to advance to the white-collar jobs that were proliferating . . . or, face diminished economic prospects" (p. 105). Empirical studies from other parts of the world, similarly, challenged the downward mobility thesis of segmented assimilation theory

(Crul & Schneider, 2009; Waters et al., 2010). Taken together, these studies questioned the basic tenets of the cumulative-progressive approach. The old assimilation canon that immigrants aspire to become more like middle-class Protestant whites in the US context was being criticized.

Another criticism of cumulative-progressive approach is from studies on ethnic enclave economy. Ethnic enclaves are local clusters of businesses that are set up by immigrants drawing labor from the immigrants of the same ethnicity. These enclaves shelter the refugee or immigrant groups from the pressures of the open market, while helping the newcomers of the ethnic community find jobs in the host country. The ethnic enclave hypothesis states that the earnings returns for the human capital of immigrants in these enclaves are comparable to the earnings returns of the immigrants in the primary sectors of the open market (Wilson & Portes, 1980). In large differentiated enclaves, the economy is often autonomous to the extent that it can sustain without any interaction with the open market outside it. In an early example where this hypothesis was tested, the economic status of Mexican and Cuban immigrants in the US was compared. While the Cuban immigrants in the enclaves of Miami enjoyed earnings returns to human capital on par with the earnings returns in the primary sector in the open market, the Mexican immigrants were less prosperous due to their history of illegal immigration and employment in secondary sectors (Portes & Bach, 1985). The concept of ethnic enclave economy itself has been criticized for being considered a homogeneous category, without differentiating between employers and employees and place of residence and place of work of ethnic immigrants (Davis, 2004). However, there is some consensus among scholars that these enclaves are intermediate stages of assimilation for the ethnic group. Here again, the cumulative progressive approach fails to explain how ethnic loyalty, as seen in the enclaves, translates to social mobility and assimilation of ethnic groups.

In the realm of communication research, the cumulative-progressive approach had an enduring impact. For Kim (1977, 1978, 1984, 2001), one of the main proponents of this approach, migrants' perceptual complexity with regard to the host society's culture represents their acculturation. If migrants have attained a nuanced and differentiated understanding of the host society culture, they are considered acculturated to it. For attaining perceptual complexity, migrants' interpersonal communication with the host society, as well as the level of access and use of mass media, needs to be qualitatively and quantitatively high.

Migrants' interpersonal communication involves most of their offline activities, for example, participation in neighborhood residential groups and recreation clubs. Interpersonal communication has a more straightforward impact on acculturation, such that those migrants who communicate with the host society more will be better acculturated than those who communicate less with the host society. Migrants' engagement with mass media, instead, has a nuanced impact on acculturation. For example, national and ethnic media including print, radio, and television, have both been identified traditionally as facilitators of acculturation, although their functional roles differ (Fathi, 1973; Zubrzycki, 1958).

Kim's (1977) theory attributes greater role for host society media in acculturation, such that those migrants who have higher exposure to such media will have higher perceptual complexity with respect to host society culture. However, another group of mass media—the ethnic media—also influence acculturation, but in a qualitatively different manner. Ethnic media is defined as those published or broadcasted from the migrant-receiving country, and those cater to the needs of respective migrant groups. Their functions include providing information to the migrant community related to education, housing, jobs, and lifestyle. Ethnic media facilitate the creation of a sense of community among the diaspora, much like mass media help create "imagined communities" based on nationalistic feelings (Anderson, 2006). Migrants rely on ethnic media in their attempt to gain information on the new country to which they have migrated, but also to retain links with their home culture. From a cumulative-progressive perspective, as migrants' sustained engagement with the ethnic media strengthens their ethnic identity, it hampers their acculturation to the host society (De Fleur & Cho, 1957; Graves, 1967; Spindler & Goldschmidt, 1952).

The acculturation theory developed by Kim (1977, 1978), rooted in communication, was followed by a number of empirical studies in different cultures, focusing on a number of factors. For instance, scholars in this school of thought asked questions like "Can the ownership of a radio or a television in the host country lead to immigrants' greater exposure to host society media?" and "Can exposure to host society media influence migrants' adaptation to the host society?" Other factors like immigrants' income, age at the time of immigration, educational background, personality characteristics such as gregariousness, tolerance for ambiguity, and familiarity with the host culture before immigration and the similarity between immigrants' home and the host cultures were also studied while evaluating the impact of communication on acculturation (Goldlust & Richmond, 1974; Kim, 1977; Yum, 1982). Assuming an "assimilationist" approach, these studies recapitulated that consumption of more of host society-oriented communication means more acculturation, whereas consumption of more of ethnic media translates to strengthened ethnic identity (Barry, 2006; Hao & Zhu, 2005; Kim, 1977).

This foundational principle of the cumulative-progressive approach has been criticized by a number of scholars. They observe that the consumption of ethnic mass media is not detrimental to acculturation. Consumption of ethnic media is not only "a positive function in a normative desire for ethnic identity" (Yaple & Korzenny, 1989, p. 312), but it also facilitates migrants' assimilation to the dominant host culture (Park, 1971; Zubrzycki, 1958). Cultural communication within the same ethnic networks helps new immigrants navigate the adaptation process and function well in the host society (Croucher, Oommen, & Steele, 2009). To be sure, consumption of mass media is not uniform across different segments of the diaspora. It is observed that with every passing generation, there has been decreasing interest among immigrants to keep their traditions alive and decreasing interest to access news from homeland. The very idea of homeland itself would correspond with the host country for later immigrant generations.

To summarize, the cumulative progressive approach, rooted in a philosophy of Anglo-conformity, had some hold in acculturation studies in its early years. More nuanced understanding of the process was achieved while studying phenomena like segmented assimilation and ethnic enclave economy. Based on these understandings, the cumulative progressive approach was criticized for being uni-dimensional in its conception of acculturation and for subscribing to an Anglo-conformist philosophy. From a communication perspective, there is yet another drawback in the cumulative progressive approach, which is that this approach does not give sufficient attention to the possible state of "regression" in migrants' relationship with the host culture. In the emerged communication technology scenario, mobile communication to home culture represents this possible state of regression, as is evident in migrants' prolonged engagement with their home culture (Vertovec, 2004). Studies have shown how the reduced cost of international phone calls via mobile phones has increased migrants' communication with their country of origin (Law & Peng, 2008; Vertovec, 2004). Diminescu (2008) observes that migrants, with the help of ICTs, increasingly maintain their remote relations as "relations of proximity and . . . activate them on a daily basis" (p. 567). Migrants' sustained relationship with the home culture also has the downside of supporting ethnic enclaves that resist integration with the host culture (e.g., the Hasidic Jews in New York). A pluralistic-typological approach gained importance in the wake of this need to incorporate migrants' engagement with multiple cultures.

The Pluralistic-Typological Approach

The pluralistic-typological approach maintains that migrants can show affinity towards their home culture, while simultaneously relating themselves to the host society (Berry, 2001). This approach has been characterized by a bi-directional model, which proposes a typology of acculturation based on two hypothetical questions:

(a) *to what extent does an in-group or out-group want to maintain its own ethnic culture; and*
(b) *to what extent does an in-group or out-group want to relate to other cultures* (Berry, 2005) (Refer to figure 2 in appendix B.).

These two dimensions are not mutually exclusive. Instead, depending on the degree to which migrants or host society members answer each question, their acculturation orientation is marked as *assimilation, integration, marginalization,* and *separation,* which are mutually exclusive categories. Those migrants who are *integrated* would be favorably disposed to both home and host cultures, whereas those who are *assimilated* would have positive attitudes towards host culture and negative attitudes towards home culture. On the negative side, those who feel completely *separated* would seclude or ghettoize themselves from members of the host culture and deliberately and exclusively associate themselves with members of the home culture, whereas those who feel rejected by both host and home cultures fall into the category of *marginalization.*

To think about the bi-dimensional model in terms of real-world situations, immigrants to the United States in the early twentieth century were mostly oriented towards assimilation. Immigrants who found their long-term residence in the host country beneficial were more than willing to forgo their cultural heritage and embrace Americanness. Integration is evident in modern day Singapore where "foreign talent" immigrants from India, China, and other countries in South and Southeast Asia are willing and are allowed to retain their ethnic cultural heritage. Despite the retention of their ethnic characteristics, these immigrants are keen on taking up permanent residency and becoming citizens of Singapore in the long term. The third category, separation, is the attitude of most low-skilled migrant groups in developed countries. They are separated mainly due to their lack of permanence in work status in the host country. These migrant workers are often disallowed options for settlement in the host country, and as a result, their engagement with the host society remains low. Persecuted artists or activists in exile can be good examples of marginalized individuals. They are seldom accepted by their own ethnic communities in the host country, but also find themselves as foreigners among members of the host society.

The pluralistic-typological approach shows greater sensitivity towards migrants who have remained "ethnic" even after their prolonged engagement with the host society. This implies that certain groups are more willing to assimilate while other groups may prefer integration. The intention, attitude, or orientation of the immigrant communities is crucial in assessing its acculturation. The question then would be "How free the immigrants are to follow the orientation they naturally have with the host society at their will?" Studies have pointed out that immigrants are more often dictated by the policies of the host government and influenced by the attitudes of the host society (Birman, Trickett, & Buchanan, 2005). Factors like group vitality and economic standing of the immigrant group also influence immigrants' freedom. Liu (2007) highlights the relationship between group vitality and migrants' freedom in choosing the acculturation orientations they want:

> The stronger the vitality of immigrant groups, the more likely such groups are to adopt [acculturation] orientations that reflect their own priorities rather than those determined by the host majority. (pp. 771–772)

Thus, a group of immigrants entering Canada or US under the investor visa programme will attract lower resistance from the civil society and the authorities on the cultural traditions they want to follow. Whereas, a low-income group like migrant workers will have to face top-down rules that check their integration to the host country.

While the two hypothetical dimensions—retention of cultural identity and establishment of contact with other cultures—are presented as migrant-oriented choices, the host society also has its own rules guiding integration. Achieving a congruence of orientations between the immigrants and the host society is paramount. Bourhis and others (1997) map host society's acculturation orientations onto migrants' orientations to predict the overall situation as consensual, conflictual, or problematic. It is based on the idea that just as immigrants orient towards the host society, the latter too adopts strategies to engage with the out-groups or the immigrants. In fact, both conflictual and problematic situations suggest a mismatch in the attitudes of immigrants and the host society. For example, following the recent flow of refugees from the war-torn countries in the Middle East and Asia to Europe, there have been reports of conflicts between refugees and the host society in a number of cities in Europe. These conflicts have resulted from a mismatch in host society's expectation and the refugees' way of deportment in the new country. The efforts are then directed towards bridging the gap between host society's expectations and immigrants' different ways of acculturation.

Cultural fusion theory makes a similar argument. It posits that just as the newcomers in a culture learn and adopt new behaviors/traits of the dominant culture and retain their cultural identities, the dominant culture too is transforming as a result of newcomers' interaction with the host society (Croucher & Kramer, 2017; Kramer, 2000). Moreover, cultural fusion is an open system, wherein the multitude of factors that affect the adaptation of migrants and the host society—individual's identity, personality, biology, and demographics—is in constant flux:

> As the newcomer and the dominant cultural group are interacting together, in an open system, it is impossible for the dominant culture to not be affected to some extent by the introduction of new cultural practices, beliefs, norms, and so forth. (Croucher & Kramer, 2017, p. 102)

An example would be the way in which immigrants from Asia, Europe, and Africa have transformed the US society in the last century.

The discussion so far has been along the dimensions of home culture (in terms of migrants' retention of cultural identity) and host culture (for example, the level of contacts maintained by immigrants with other cultures and the host culture's reciprocity). Recent research has situated migrants beyond the boundaries of sending and receiving countries, and identified them in a larger political context. Burrell and Anderson (2008) introduce the idea of an "aspirational migrant" who temporarily breaks free from his or her own culture to explore the possibilities offered by globalized spaces. This concept identifies migrants as being driven by emerging opportunities in the globalized world. Their movements are not limited to those in the country of origin and destination. An aspirational migrant's residence in any country may be considered as "only one point in a series of moves or travel plans" (Burrell & Anderson, 2008, p. 207). Another concept that helps locate migrants beyond the binary of home and host country is "transnationalism." In the era of digital communication, this concept has increased relevance, as it captures migrants' activities and exchanges with a global reach, beyond the boundaries of country of origin and destination.

Transnationalism

The pluralistic-typological approach has been enriched by the concept of transnationalism, widely used in contemporary studies on immigrants. Transnationalism refers to migrants' cross-border activities and imaginings including remittances, political engagement, and cross-border activism. Transnationalism explains the intricate cross-border connections and movements of migrants who are driven as much by economic pressures as by career aspirations, family obligations, or any emerging demands of interconnected lives in the modern world. We restrict ourselves to elaborating only on two domains of transnationalism below, the economic and the political domains.

In the economic domain, concepts like "ethnic entrepreneurship," "economic transnationalism," and "transnational enterprises" are used to denote these cross-border engagements. Although these concepts denote different phenomena, what they have in common is their association with economic activities of immigrants with transnational implications (Zhou, 2004). Cross-border economic activities of immigrants are varied. At a very basic level, it represents a flow of goods, services, or money across borders. Take for example, a grocery store set up by an immigrant to cater mainly to the members of his/her ethnic community in the host country. He or she will be sourcing most of the products from the home country. The price differential between the home and host countries, less the operational cost, is the profit. Most often, this profit is routed back to the immigrant's country of origin or reinvested in the expansion of the business in the host country itself.

Ethnic businesses range from capital and skill-intensive ventures to retail selling of goods and personal services (Landolt, 2001; Ley, 2006). Apart from capital, labor, and land, information is crucial to plan and set up such businesses. The potential entrepreneur would need information on the business climate, the types of enterprises that are viable among the ethnic community, and the regulatory regime of the host country. Digital technologies have helped immigrants find this information and have offered solutions to problems faced by immigrant businesses. Ethnic businesses use social media platforms for brand building, while ethnic members advertise their traditional artisanal skills and sell ethnic products on online platforms.[1]

The reader may note in passing that a parallel development of business can be observed in migrants' home countries, assisted by the remittances. Apart from direct investments, the diaspora invests in businesses indirectly through company shares and partnerships. A large part of these remittances are used in developmental works, while some portion is channeled to human capital development (e.g., education). A large body of literature examines how the remittances of immigrants to their home countries have bolstered the village and urban economies in terms of increased investments in agriculture and trade (Mendola, 2008; Rahman et al., 2014). Studies have reported how migrants actively support the causes of nation-building and make financial contributions for social welfare activities at local and global levels (Bernal, 2006; Parham, 2004; Velayutham & Wise, 2005). The ease provided by digital banking methods for the repatriation of funds is key to this cross-border economic activity. Such economic activities also result in increased cross-border communication, supported by ICTs and smartphones.

Returning to our discussion on migrants' economic activities in host countries with transnational implications, the social capital of immigrant businesses is an important factor that determines their success. However, different classes of migrants have different needs for social capital. Low profile immigrant entrepreneurship relies on social capital more than big investments.

In countries like Australia, Canada, Singapore, and the US, big investors are given opportunities to naturalize to the host country, opening up the possibilities to extend their market to other sections in the resident population. This reduces their reliance of the social networks within the ethnic group. Big investors also have other advantages (e.g., better education and management experience) when they arrive in the new country (Marger, 2001). Low-key enterprises, on the other hand, are reliant on ethnic and family networks in the host country. These enterprises generally employ family members and others from the ethnic social network in order to reduce costs. Their businesses also cater to the "captive market," at least initially, although in many cases they later expand to a broader clientele (Rath & Kloosterman, 2000). A study conducted by Glover and Sim (2002) identified the role of social networks in the success of ethnic businesses. While weak economic and market conditions were the most influential factors in the failure of immigrant enterprises, limited social networks and connections of the entrepreneurs contributed to their failure (Glover & Sim, 2002).

In sum, in the immigration context, low-profile businesses and low-income groups are more reliant on social networks and social capital for their very livelihood. Big immigrant investors rely more on human capital than social capital. Digital media have become popular platforms for all types of immigrant entrepreneurs to promote their businesses. This resonates broadly with the research findings among low-income groups in the developing countries, who streamline their work processes using mobile communication, often tapping into social networks. Finally, if one is to make a connection between ICT diffusion and the rate of change in economic transnationalism, a paradoxical situation is likely to emerge. On the one hand, there is an expansion of the international capital, avenues for investments across borders, and a widening scope for diversification of the ethnic enclave economy. On the other hand, the social structures of the same ethnic enclaves are strengthened by drawing material support from the ethnic businesses that flourish transnationally.

In the political domain, creation of a transnational space denotes diffusion of political ideals across borders (Garrett et al., 2012), establishment of transnational activist networks (Keck & Sikkink, 1998), migrants' participation in nation building using online platforms (Alinejad, 2011; Laguerre, 2005), and the creation of a global public sphere (Volkmer, 2014). In political transnationalism, the role of the state has been conceptualized in rather opposing ways, which has led to incongruent conclusions on migrants' role in the phenomenon. According to one position, state's powers among the migrant population have weakened due to virtual networks across borders and the emergence of a global public sphere that upholds secular, democratic ideals. Migrants help disseminate ideas and beliefs to their homeland resulting in decisive changes in its politics and helping in nation building: "the information that migrants channel to high-volume migration communities has an aggregate-level effect that alters attitudes and beliefs of members of those communities" (Pérez-Armendáriz & Crow, 2010, p. 120). This loosening grip of the state among migrants has been utilized in proposing a criticism against "methodological nationalism." Wimmer and Glick Schiller (2002) argued that "methodological nationalism" provides an insufficient framework for studying migration, as it conceives migrants as part of "a bounded national container society" (p. 324). In order to move beyond the limitations of methodological nationalism, it is important to adopt categories that are more fluid and non-essential. Transnationalism has been proposed as one such concept that can overcome the boundedness of methodological nationalism.

The state has been accorded greater primacy in an alternative position on transnationalism. Despite all apparent weakening of their powers among the migrant population "states reconfigure themselves and redefine national membership to maintain ties to and profit from their transnational constituencies" (Levitt, DeWind, & Vertovec, 2003, p. 568). This has the consequence that the power structures, as determined by the state, but also based on political and economic considerations, will remain in place, even though virtual networks may appear to be transparent and egalitarian. The discrimination based on caste, class, race, and gender that migrants face in the offline world may also continue to be reflected in online networks.

The political power differentials between migrants' country of origin and destination also interact in complex ways between migrant statuses to decide on the cross-border political participation of migrants. For example, Latin American and Caribbean migrant workers in the US are found to be influencing the politics of their homeland in substantial ways (Barry, 2006; Laguerre, 2005; Parham, 2004). Whereas, the influence of Indian or Chinese diaspora on the politics of their homeland is comparatively low.[2] The class differentials in the political influence of diaspora also need to be considered. Low-skilled transient workers have limited entitlements in the political processes of their home countries due to their low economic and political standing. Low-skilled migrants from democratic states like India do not appeal to the political parties as an important constituency. Thus, while it is normal for political parties in India to woo investment from Indian-origin expatriates and diaspora the world over, the parties rarely campaign among low-skilled migrants in Singapore or the Middle East. The restrictive political space in the migrant-receiving countries exacerbates the problem of low-skilled workers' lack of access to the public sphere.

New media platforms, especially mobile phones, have been found to be helping the development of a public sphere, where migrants can express their opinion about political issues relating to their home country. As a result, they co-create their identities in virtual spaces. However, the accessibility to such a public sphere is not evenly distributed among all segments of the migrant population; expatriates who have access to the public sphere respond more readily than low-skilled migrant workers (Suroor, 2011). Such political participation is also restricted in countries that allow limited freedom of expression, like Singapore and the Middle East. Virtual spaces can be both a less-controlled avenue and a space of constraint for migrants to negotiate their political belonging and identity, while locating their day-to-day lives in the larger political context of home and host countries.

The concept of transnationalism, owing to its ability to evaluate the networks that migrants develop, can take different dimensions ranging from economic to political, and cover a range of cross-border activities including investments, activism, and familial communication. The concept helps overcome the boundaries of home and host countries and situate modern day migrants in a broader socioeconomic and political context. Transnationalism is, thus, a concept that can help strengthen the pluralistic-typological approach.

In sum, the pluralistic-typological approach explains heterogeneity in social and cultural processes better than the cumulative-progressive approach. It is also less normative towards the acculturation phenomenon since the two axes along which it identifies the four types of acculturation (*assimilation, integration, marginalization,* and *separation*) allow greater freedom for less-influential migrant groups to hold on to their preferred acculturation strategies. Expanding the possibilities of the acculturation models, studies have also attempted to link acculturation with mobile phone appropriation, identifying typologies as well as links between the two phenomena (Aricat, Karnowski, & Chib, 2015). This book further advances the knowledge by exploring the processes that can explain the links between mobile phone appropriation and acculturation.

We defined the variables pertaining to mobile communication in the last chapter. Specifically we looked at mobile phone communication to dominant cultures in host society and mobile phone communication to migrants' home culture. In the remaining part of this chapter, we explain variables representing acculturation and adaptation. Due to low construct validity of variables on acculturation orientations—assimilation, integration, marginalization, and separation—we had to use two proxy variables—*cultural identity* and *multiculturalism in workplace*—in the study. A detailed discussion on the reasons of low construct validity of these variables is included later in the chapter. The nature and direction of the relationships between the two proxy variables representing acculturation and migrant adaptation are also discussed in the following sections.

Cultural Identity

Cultural identity[3] denotes the strength of loyalty and extant affinity of migrants towards their home culture. Cultural identity often overlaps with concepts like ethnicity and nationalism. Two kinds of nationalism are generally identified: political and cultural nationalism. Cultural identity can be a trigger for cultural nationalism. Hutchinson (1994) observes that cultural nationalism is akin to attributing organic life force into the tradition, history, and culture of a nation. Cultural nationalists identify state institutions as part of an organic unit, which requires regeneration at the hands of each member of the society. Incessant rediscoveries of the authenticity of one's culture happen as members search for meaning and essence in "a set of repetitive mythic patterns, containing a migration story, a founding myth, a golden age of cultural splendor, a period of inner decay and a promise of regeneration" (Hutchinson, 1994, p. 123). Political nationalism, which is tangential to our discussion here, identifies state institutions as resting on rational foundations.

Cultural identification denotes the rootedness and the strong loyalty one feels towards one's culture. The counter concept is that of "marginal man," introduced by Park (1928), as someone who has mentally reached beyond the traditional boundaries of his/her own culture as a result of interaction with other cultures, either out of volition or by external force. According to Park (1928), marginal man, a stage he identifies as more mixed than a "cultural hybrid," alone can galvanize societal change and progress. Marginal man, unlike a patriot, is capable to shake off the status quo that has afflicted settled societies. In this interpretation, "marginality" may be antithetical to notions of well-being in any given point in time,[4] but it alone ensures progress of societies in the long-term. Stonequist (1935) elaborates on "marginal man" and identifies that marginality "represents a process of abstraction, a core of psychological traits which are the inner correlates of the dual pattern of social conflict and identification" (p. 10).

Among immigrants, cultural identity can switch between states of dormancy and activation. Rumbaut (2008) introduces the term "reactive ethnicity" to denote cultural identity that is evoked in the face of host society's discriminatory practices and policies. Identified mostly in second-generation migrants, "reactive ethnicity" denotes "one mode of ethnic identity formation [among ethnic minorities] that highlights the role of a hostile context of reception in accounting for the rise rather than the erosion of ethnicity" (p. 110). Among migrant workers in Singapore, an activation of ethnic/cultural identity was observed when a Tamil-Indian worker was killed in a road mishap in 2013. The mob reacted violently when it sensed a kind of discrimination at the hands of the emergency response team towards the community.

To further contextualize the idea of reactive ethnicity, events like 9/11 in the United States triggered a kind of fear of identity even among the well-integrated South Asians in the country. They felt that the cultural symbols with which they had been associated, as well as their racial similarities with the actual perpetrators of the violence, would trigger a backlash against them (Bhatia & Ram, 2009). Instances such as these show that cultural identity of immigrants is co-constructed along with the host society based on the former's experience in the host country (Croucher & Kramer, 2017).

Previous research has shown mixed results while examining the link between ethnic identity and acculturation (Jeffres & Hur, 1981). A strong sense of cultural identity, without sufficient identification towards the larger host society, can result in low adaptation to the host society (Nazroo & Karlsen, 2003; Ryan, 2010; Sam, 2000). This can be the result of government policies that hinder integration of all sections of the immigrant community into the society. For example, Rahman and Kiong (2013) identify a differential exclusion model followed by Singapore wherein "migrants are integrated temporarily into certain areas of society (mainly the labor market) but excluded from others such as welfare systems, political participation and national culture" (p. 86).

Similarly, migrants with integrated ethnic identity will have strong identification with their country of origin, but they would also identify with the host culture, simultaneously (Hernandez, 2009; Jasinskaja-Lahti & Liebkind, 1999; Phinney et al., 2001). Strong co-national identification has been associated negatively with psychological well-being of migrants, whereas strong host national identification leads to better sociocultural adaptation (Ward & Rana-Deuba, 1999). A strong sense of loyalty among workers towards members of the same ethnic and racial group can result in workers' low commitment towards an organization that has a multicultural character (Fitzsimmons, 2013). Since the extant literature largely supports a negative relationship between cultural identity and migrants' adaptation to both social and work domains, the related hypotheses included later in the chapter reflect this trend.

This section reviewed the factor of cultural identity, a variable that is considered as a proxy variable to acculturation. It identified the nature and direction of the relationship between cultural identity and migrants' adaptation to social and work domains, as reported in the literature. We now discuss a factor that can potentially affect migrants' adaptation in social and work domains (i.e., their openness towards a multicultural work team).

Multiculturalism in Workplace

Multiculturalism in workplace denotes the level of acceptance and willingness for co-existence a migrant worker displays towards other cultures and coworkers. This concept has relevance in migration studies as integration of international work force to the organization is a major challenge for the Human Resources personnel. This integration to workplace has broader implication, as it decides the employee's integration to the society as a whole. The study of multiculturalism in workplace is part of "diversity" studies that are based on organizations. Ely and Thomas (2001) classify diversity in workplace into three categories: (i) diversity in identity group memberships, such as race and sex; (ii) organizational group memberships, such as hierarchical position or organizational function; and (iii) diversity based on individual characteristics, such as idiosyncratic attitudes, values, and preferences. The impact of diversity in organizations on the functioning and productivity of groups has received wide research attention. Our interest is on individual worker's perception and attitude towards cultural diversity in the organization.

Diversity is perceived as racial and socio-cultural differences such as language, food preferences, and shared meanings (Larkey, 1996; Tsui, Egan, & O'Reilly, 1992). However, each individual expression of diversity is perceived in the context of the larger group that it represents and the social and economic standing of the group in a broader world. Prior literature shows that the power positions of different racial and social groups in the outside world are also reflected in the organizational context (Griffith et al., 2007; Hood & Koberg, 1994).

Ely and Thomas (2001) expanded the concept of "diversity perspective" as a factor moderating the impact of culturally diverse groups on work effectiveness. "Diversity perspective" denotes "'group members' normative beliefs and expectations about cultural diversity and its role in their work group . . . and beliefs about what constitutes progress toward the ideal multicultural work group" (Ely & Thomas, 2001, p. 234). Although earlier research found a normatively negative influence of multiculturalism on individual adaptation, leading to conflict and confusion (Griffith et al., 2007), recent research has identified positive outcomes. For example, multiculturalism is found to be positively influencing cognitive complexity (Tadmor, Tetlock, & Peng, 2009), adaptability (Chae & Foley, 2010; Chatman et al., 1998; Frankenberg et al., 2013), and opportunity for integration and learning (Ely & Thomas, 2001).

Studies have attempted to link acculturation attitudes in social life with job-related outcomes, like affective workgroup commitment[5] (see, for example, Lu, Samaratunge, & Härtel, 2012, 2013). This implies that those immigrants who have a good relationship with the host society in general will naturally feel an emotional attachment towards their work organization. Most of these studies begin with an exploration of the social adaptation of immigrants before moving to assess the job-related outcomes. This premise of the social couching the "work-related" is based on the characteristics of the immigrants under study. Most of them are likely to settle in the host society permanently, and are active members of the neighborhood organization. This premise may not be true in the case of temporary migrants, whose adaptation is largely determined by their relationships in the work organization. To fill this gap, this study explores how multiculturalism of migrants in the workplace determines their adaptation not only to work but also to the social life domain in general. As suggested by the review, the nature and direction of relationship between multiculturalism in the workplace and adaptation in social and work domains have been indicated in the hypotheses stated.

The Bi-Dimensional Model: A Critique

Before we move on to explain the adaptation outcomes, a few critical comments on the bi-dimensional model of acculturation are in order. This justifies our adoption of proxy variables "cultural identity" and "multiculturalism in workplace" in the study. The bi-dimensional model, the reader will recall, worked with a 2x2 matrix, with two hypothetical questions describing four conditions, or acculturation strategies. These strategies are easy to define, but not to operationalize in a survey. Scholars have criticized them for being overlapping, ipsative,[6] double-barreled, and non-falsifiable (Hicks, 1970; Rudmin, 2003). Ward and Kennedy (1994) argue that the "quadri-modal approach to acculturation attitudes has essentially ignored the direct and independent examination of the more fundamental components of acculturation—cultural identity and intergroup relations" (p. 333). "Quadri-modal" refers to the four-type acculturation typology mentioned above, from which one type would emerge as the dominant acculturation strategy for an individual migrant. Ward and Rana-Deuba (1999) criticized the four-fold paradigm stating that when the model is operationalized, each acculturation type is identified using "double-barreled" questions (e.g., questions that touch upon more than one issue, triggering an ambivalent meaning in the respondent [Fowler, 1992]).

Scholars have also criticized the ipsative or forced choice nature of the constructs that represent migrants' affinity to home and host cultures and the questions used to measure the four acculturation attitudes (see, for example, Rudmin & Ahmadzadeh, 2001; Sayegh & Lasry, 1993). In order to overcome the problems, Rudmin and Ahmadzadeh (2001) suggest orthogonal or direct and independent measures of attitudes. This means that the respondents should be given more direct options to choose from. For example, items like "I am proud of being a [ethnicity] immigrant" and "I would like to hang out with ethnic friends" represent more direct statements. Following this, another set of direct questions related to the host society may be included (e.g., "I desire not to have friends among the local population of this country" and

"I do not want to learn the language of the dominant community in the country"). In technical terms, these two sets of questions on two different cultures are "orthogonal." There is no ambiguity in interpreting the responses. Whereas, in the current operationalization of the quadri-modal model, a typical question combines and confounds the dimensionality of cultures. For example, a question like "It is important for me to be fluent in both [ethnic language] as well as the languages used in [the host country]." A response to this can create problems in statistical analysis.

This led us to adopt singular concepts (independent constructs, as opposed to compound constructs like *assimilation* or *marginalization*) like *cultural identity* and *multiculturalism in workplace* in the present study. The pluralistic outlook of bi-directional model is retained, while making the question more direct on the cultural affinity and attitude of the respondents. The following sections in the chapter discuss the adaptation outcomes, with specific focus on two domains—social life and work.

ADAPTATION OUTCOMES IN SOCIAL AND WORK DOMAINS

So far the attention has been largely on acculturation, with greater emphasis on orientations, strategies, or attitudes. A dimension that requires greater attention is the host society's reciprocity towards migrants' attitude towards acculturation. It can be better studied under different domains of acculturation, for example, life and work domains, which are more relevant to the present research. The acculturation profile of a migrant is defined as a confluence of different domains, that is, political and government systems, labor and work, family, and religious beliefs and custom (Navas et al., 2005). For each domain, migrants use separate strategies for acculturation, which ultimately decide their adaptation outcomes (Mendoza, 1984). Migrants, for example, often retain their religious beliefs and practices despite moving to a new country. Whereas, there are major changes observed in their work practices after migration. Scholars identify areas like religious beliefs as "hard core," where adjustments are difficult to attain. Rather it is domains such as work life and economic practices that are the "peripheries" in which adjustments are more readily achieved (Miller et al., 2013).

The focus on the reciprocity of host society as well as domain-wise attention on acculturation leads to the questions: "How does acculturation attitude help decide adaptation outcomes of migrants in different life domains?" and more pertinently "How does mobile phone communication lead to migrant adaptation in different domains of life." This section reviews the outcome factors used in the study, that is, the level of migrant adaptation to life and work domains. While adaptation to social life has been characterized by three sub-constructs—*proficiency in host society language, adaptation to host society rules,* and *adaptation to host society's custom and lifestyle*—adaptation to work life has been characterized by the factor *organizational commitment.*

In order to guide the inquiry, the first focus of this analysis is: *How do cultural identity, multiculturalism in workplace, and mobile calling to home and host cultures affect migrants' adaptation outcomes in the host country?* The factors introduced in the first part of this question have been reviewed in previous sections. The second part, pertaining to adaptation to social life, is reviewed in the following section.

The Social Domain

Three factors will be used to evaluate the acculturation outcomes of migrants in social domain. These are (1) proficiency in host society's language, (2) adaptation to host society's rules, and (3) adaptation to host society's customs and lifestyle. Combined, these factors determine the social adaptation of migrants.

Proficiency in Host Society's Language

Gaining proficiency in host society language is considered a necessary prerequisite for optimal adaptation of migrants in the new culture. Language proficiency is also a measurable index of migrant adaptation. Scholars have acknowledged that individuals who look beyond their cultural identity and preferences learn new languages quickly and adapt easily to the cultures to which they migrate. Citron (1995, p. 106) notes: "A language learner who can empathize with his or her interlocutors and recognize the culture-boundedness of each language would be better able to learn the new language than one who cannot." By learning the language of the host society, a migrant can build a complex and nuanced perceptual structure of the host culture (Kim, 1977). This perceptual complexity has been found to have a positive relationship with the migrant's knowledge of and adaptation to the host society. Knowledge of host society language reduces acculturation stress leading to better adaptation: "To the extent that immigrants possess the required background in understanding and using the language of the host culture . . . their acculturation process is likely to be less stressful" (Krishnan & Berry, 1992, p. 193).

Singapore presents a complex picture on languages used in business and more generally in society. There are four official languages, English, Chinese, Malay, and Tamil. Large-scale migrations from China and the southern coast of India in the nineteenth and early twentieth centuries resulted in long-term settlement of Tamil- and Chinese-speaking populations in the country. British colonialism and the nation's legacy of political and social attachment with Malaysia resulted in the inclusion of English and Malay in the official language list. For migrant workers from South Asia, the most commonly shared languages with the resident population are English and Tamil. The latter is, however, specific to people migrating from one particular state in India, which is Tamil Nadu. Due to cultural and geographical proximity, migrants from neighboring states of Tamil Nadu, like Kerala and erstwhile-Andhra Pradesh,[7] may be proficient in Tamil to a limited extent. For Bangladeshis, another South Asian group focused on in this study, the most accessible language remains to be English. Since English is a second language in Bangladesh and India, and since its accessibility is mostly limited to the educated classes, migrant workers at the low-income categories usually find it a major barrier in their job mobility and cultural adaptation. This situation is further exacerbated by the fact that Mandarin and other Chinese dialects are the first languages of a large percentage of residents, and sometimes the only language known to older generation Singaporeans (Chua, 2003). Despite this, most migrants do not have to learn Chinese, but mastery in English, at least at a functional level, is important for their survival. This study evaluated migrant's competency in English, a language more relevant to their situation.

Adaptation to Host Society's Rules

Asian political contexts and governance styles are so diverse that the continent is home to democratic and autocratic states alike. Liberal and conservative groups thrive in varying numbers in this region. Such diversity in the culture, governance, and politics of countries in Asia suggest that even the South-South migrants have to adapt to the host society. In Singapore, the state clearly demarcates between different classes of migrants when rules are enforced, posing greater control on the movements of migrants at the lower strata (Hamid, 2015). The extent to which migrants adapt to the general rules of the host country decides their adaptive behavior in social life. Scholars observe that adaptation involves a kind of self-disciplining by the workers, who are subjected to strict schedules, instructions, and monotonous life patterns, which would eventually lead them to become "good, docile *Others*," as required by the host society (Abdullah, 2005, p. 232). The importance of low-skilled immigrants' compliance to Singapore's rules and social norms has been reiterated by authorities in the aftermath of December 2013 riots in Singapore (MHA, 2014; PTI, 2014). Thus, the set of laws and rules (e.g., law prohibiting drug trafficking and littering, in the case of Singapore) becomes a negotiation platform for migrants in their cultural adaptation process, especially if they have taken up assignments in different countries in Asia before moving in to Singapore.

Adaptation to Host Society's Custom and Lifestyle

Migrants often complain about some unknown discomfort that they feel in the new culture. Skuza (2007) notes that there is an increasing tendency in acculturation studies to capture feelings of personal and social isolation and confusion experienced by migrants to measurable psychological categories of depression. However, the emphasis should be given to the overall comfort level of the migrants rather than compartmentalizing of the experiences (Skuza, 2007). Mutual Intercultural Relations in Plural Societies, an acculturation evaluation project based on bi-directional model, has incorporated this concept in its theoretical schema (Berry, 2005). This study adapted the concept of adaptation to host society's custom and lifestyle to suit the migrant experiences in Singapore. Ranging from one's opinion about the status of workers in the host society to, for example, one's opinion about the nightlife typical of metropolitan culture, the construct "adaptation to host society's custom and lifestyle" intended to capture migrants' mental and physical comfort in host country Singapore.

Based on the discussion of mobile communication to home and host cultures, cultural identity, multiculturalism in workplace, and adaptation to social life, the following hypotheses are formed:

H1: Multiculturalism in workplace, mobile phone calling to dominant host cultures and mobile phone calling to home culture positively affect migrants' adaptation to social life, whereas cultural identity negatively affects migrants' adaptation to social life.

H1a: Multiculturalism in workplace positively affects adaptation to social life.

H1b: Mobile phone calling to dominant host cultures positively affects migrants' adaptation to social life.

H1c: Mobile phone calling to home culture positively affects migrants' adaptation to social life.

H1d: Cultural identity negatively affects adaptation to social life.

The Work Domain

While there have been studies in different cultural contexts focused on evaluating the acculturation orientations of migrants in social life (e.g., Dandy & Pe-Pua, 2010; Horenczyk & Munayer, 2007; Liu, 2007; Piontkowski et al., 2000), management scholars acknowledge that "the concept of acculturation has rarely been discussed in the context of organizational behavior" (Cox & Finley-Nickelson, 1991, p. 91). The study reported here used organizational commitment as an index to evaluate migrant workers' level of adaptation to work (Chan & Qiu, 2011; Claus, Lungu, & Bhattacharjee, 2011).

Organizational Commitment

Organizational commitment is a complex concept with broad definitions in literature: it is sometimes defined as antecedents and consequences of employees' attachment to an organization, whereas defined in other cases as the process of attachment or the state of attachment itself (O'Reilly & Chatman, 1986). A central theme that underpins the definitions of organizational commitment is the employees' psychological attachment to an organization. Drawing from Kelman's (1958) taxonomy of attitude change, O'Reilly and Chatman (1986) defined organizational commitment using three independent constructs—compliance, identification, and internalization. Compliance has a materialistic interpretation. It occurs when attitudes and behaviors are adopted not because an employee fully believes in them, but simply to gain specific rewards. Identification occurs when employees feel proud to be part of a firm, without however adopting its values as their own. Internalization is when employees' behavior is characterized by a congruence of their own values with that of the organization (O'Reilly & Chatman, 1986; O'Reilly, Chatman & Caldwell, 1991). The present study adopts this psychology-based definition, as it is found to be more relevant in the case of migrant workers who have no financial investments or other extrinsic factors to determine their organizational commitment.

Doubts can be raised on why "organizational commitment" is chosen as the job-outcome to be studied, overriding more popular concepts like job satisfaction or productivity. One justification is that "organizational commitment" is an intermediate stage in determining "participation, power, teamwork, and professionalism" (Chan & Qiu, 2011, p. 1113), and thus a pointer to popular terms like job satisfaction and productivity. Further, organizational commitment is a more general and stable index of employee's relationship with the organization in comparison to job satisfaction. As Mowday and others (1979) argue, "Although day-to-day events in the workplace may affect an employee's level of job satisfaction, such transitory events should not cause an employee to seriously reevaluate his or her attachment to the overall organization" (p. 226). This makes organizational commitment a somewhat more stable index over time than job satisfaction.

It is also important to explore how the links between communication and job-outcomes are explicated in literature. Research in business communication has highlighted that internal communication helps achieve better employee engagement (Mishra, Boynton, & Mishra, 2014). Saks (2006) maintained that clear and sustained communication with employees helps achieve employee engagement, which in turn results in employees' having better relationships with the employer. Communication within organization helps workers to be informed about company policies, to obtain information on professional tasks, and to solve problems (Anderson & Martin, 1995; Mathieu & Zajac, 1990; McMillan & Lopez, 2001; Rupert et al., 2010). Based on this understanding, we predict that a greater amount of communication among coworkers would be positively related to a greater organizational commitment. Conversely, migrants who still have a high level of contact with their families and others in their country of origin are less likely to be integrated to the social and work life of the host country.

The discussion on multiculturalism in the workplace suggested that migrants' open disposition to cultural diversity in the organization can produce positive job-outcomes like better organizational commitment. Conversely, greater loyalty to one's own ethnicity and race would lead to lower commitment to organization. The following hypotheses are formed in order to guide the enquiry on the effect of acculturation and mobile communication factors on organizational commitment:

> H2: Multiculturalism in workplace and mobile phone calling to dominant host cultures positively affect migrants' organizational commitment, whereas cultural identity and mobile phone calling to home culture negatively affect organizational commitment.
>
> H2a: Multiculturalism in workplace positively affects organizational commitment.
>
> H2b: Mobile phone calling to dominant host cultures positively affects migrants' organizational commitment.
>
> H2c: Mobile phone calling to home culture negatively affects migrants' organizational commitment.
>
> H2d: Cultural identity negatively affects migrants' organizational commitment.

Figure 4 in appendix B shows the relationships stated in hypotheses 1 and 2. To summarize, the chapter so far highlighted the approaches of cultural adaptation and transnationalism in the context of migrants' communication technology use and identified the variables that are relevant to the present study. It also identified relationships between factors as supported by literature and represented the relationships in the hypotheses. The appropriation approach discussed in chapter 3 was extended to this chapter to identity the acculturation typology and the related variables. However, the typology has been criticized on several fronts: First, it resists operationalization even in statistical terms. Second, as argued by diaspora studies, the typology presents a closure to human experience, as it presents migrant integration in essentialized watertight categories. The following section discusses acculturation from the perspective of diaspora studies, and based on non-essential categories. This supplements the appropriation approach discussed in chapter 3 in the context of mobile phones. The appropriation approach along with the concept of non-essential categories of acculturation provides a basis for the qualitative exploration in this volume.

ACCULTURATION AND MOBILE PHONE APPROPRIATION: NON-ESSENTIAL PATTERNS

In this section, we briefly review the studies in mobile communication and acculturation, which presented typologies, and we explain their drawbacks while explaining methods to overcome them. In mobile communication literature, quantitative studies on adoption present typologies in order to categorize users based on adoption (or non-adoption) and amount of technology used. For example, Vanden Abeele and others (2014) correlated lifestyle with mobile phone use and identified three clusters of users: *trendy users*, *thrifty users,* and *engaged users*. Each cluster significantly differed from the other on socio-demographics, use of mobile phone in the presentation of the Self, and mobile phone's role in friendship relationships. Donner (2006) studied mobile phone adoption in Rwanda based on the themes of "overall telephone scarcity" and "increasing mobile phone use," categorizing the respondents based on their adoption and non-adoption (p. 18). Similar categorization of mobile phone users based on purchase decisions, personality, and use can be found in studies across cultures (see, for example, Kim & Park, 2013; Lepp et al., 2015). Similarly, Ciunova-Shuleska and others (2012) identified clusters like "involved users" and "indifferent users" based on significant differences related to dependency on mobile phone, concern towards negative aspects and safety features of the phone, fashion-oriented uses and users' interest towards value added service on the phone. As these examples suggest, categorizing or clustering individuals based on one or more characteristics has been a practice widely used in mobile communication and internet research (Bakardjieva, 2006; Ciunova-Shuleska, Grishin, & Palamidovska, 2012; Donner, 2006; Haverila, 2011).

On the acculturation side, studies have identified clusters of immigrants based on the typology discussed in this chapter. Broadly, these clusters are formed based on the immigrants' attitudes towards the host society and their extant relationship with the home culture. These studies have focused on different countries, factoring in the external conditions like immigration policies of the host government. Although this approach of categorizing users based on quantitative data aims at making accurate depictions of reality, it has major drawbacks. First, the typologies are watertight and have hardly any room for fluid identities. Second, the typologies are based on quantitative measures, which misses out the nuanced picture of co-constructed meanings in individual life world. We introduce an appropriation approach as a way to overcome the drawbacks of typologies in mobile phone studies, and follow the approach of diaspora studies in order to overcomes the drawbacks of acculturation typologies.

A major achievement of the appropriation approach was that it had less emphasis on the functional path and concentrated on a more descriptive understanding of the phenomenon (refer to the discussion of the circular model). Added emphasis was given to the symbolic usage of the mobile phone, mainly with implications for the prestige and social identity of the user. This latter has defied tendencies to make universalist observations on the mobile usage phenomenon and as a result, helped identify non-essential patterns of mobile phone appropriation. Similar to this trend, diaspora studies defied the universal categories as well as the permanence of underlying acculturation processes highlighted in the bi-directional model. This was made possible by challenging the idea of an authentic, continuous, and progressively unfolding self that was at the core of traditional conceptualization of acculturation (Hall, 1991). An increased sensitivity towards the socio-historical process that determined the acculturation experience of migrants, especially those from once-colonized countries, characterized diaspora studies:

> [The acculturation experiences] are constructed through a dynamic, back-and-forth play concurrently between structure and self, being privileged and marginalized and is caught in the web [of] socio-political and historical forces. (Bhatia & Ram, 2009, p. 147)

This conceptualization informed the present study as it attempted to look beyond the acculturation typologies, and analyze migrants' experience beyond the two hypothetical questions of relationships with host and home cultures. The basic structures of cultural affinity and detachment that supported the typology were considered merely the starting point. Similarly, the appropriation approach helped identify mobile phone users based on non-conforming, subversive, and unique patterns of use determined by migrant conditions. This part of the study is guided by the question: *How do acculturation types relate to mobile phone appropriation types among migrants, when defined in non-essential and non-universal terms?*

An integrated diagram that summarizes the discussion on mobile phone use and migrant acculturation is seen in figure 3 in appendix B. The figure shows how the component "usage and handling" of the circular model draws on from the tradition of mobile communication studies that followed a functional approach towards mobile phone usage. It also highlights the "prestige and social identity" component of the model as drawing on from the appropriation approach followed by mobile communication studies. The integrated research design then combines the functional approach with acculturation conceptualized in terms of essential categories (the bi-dimensional model in figure 2 of appendix B), and the appropriation approach with acculturation conceptualized in terms of non-essential categories. This leads to the integration of two methods—the quantitative and the qualitative methods—used in the study.

SUMMARY

This chapter built on the material developed in chapter 3 and extended the two ways of studying mobile phone use—functional and appropriation approach—to the realm of acculturation. Parallels were drawn between the functional approach in mobile phone studies and the trends of studying acculturation by "essentializing" migrant identities. Two prominent lines of thought in such an approach—cumulative-progressive and pluralistic-typological—were reviewed. The studies have variedly identified the impact of ethnic media on the acculturation of migrants, depending on the approach they have followed and the interpretative schema they have used. Although the bi-directional acculturation model was adopted to describe the cultural affinity of the migrants with their home culture as well as to dominant host society cultures, operationalization of the constructs was not obvious among the people studied in this work. This substantiated many criticisms raised against the model by other researchers. Subsequently, constructs "cultural identity" and "multiculturalism in workplace" were adopted to measure the acculturation attitude of migrants. The two constructs were reviewed, and the nature and direction of their relationship with adaptation outcomes were discussed.

The chapter also discussed the importance of studying acculturation in different domains of migrant life as each domain demanded different communication behavior from a migrant. The focus of this research was, however, narrowed down to two domains—lives and livelihood or personal/social and professional. Although there existed overlaps between these domains, in view of clarity for empirical research this study considered them as separate.

Parallels were drawn between the appropriation approach in mobile phone studies and the trends of studying acculturation from the perspective of non-essential conceptualization of self and by ascribing greater importance to dynamic socio-historical processes. This guides the second broad area of inquiry in this volume.

NOTES

1. This social media group run by a section of Indians in Singapore provides information ranging from private tuitions to ethnic handicrafts: https://www.facebook.com/groups/122805374444657/?ref=group_header&view=permalink&id=1157327427659108

2. This needs some qualification, as the situation is fast changing in today's scenario. Major political parties in India, for instance, are using new communication technologies to seek support from overseas citizens.

3. Cultural identity has been interchangeably used with ethnic identity. For instance, Eriksen (1996) argues that ethnic identity is oppositional, as it is evoked in an individual when he or she realizes that he or she is culturally different from others. Hall (1996) identifies the role of culture in situating all knowledge of an individual: "Ethnicity acknowledges the place of history, language and culture in the construction of subjectivity and identity" (p. 162).

4. As in, for example, Durkheim's (1951) notion of anomie.

5. Apart from affective workgroup commitment, other two dimensions are continuance and normative. Affective commitment denotes employee's emotional attachment to the organization. Continuance commitment denotes awareness of the costs associated with leaving the organization. Normative commitment denotes a feeling of obligation towards the organization (Meyer & Allen, 1991).

6. Ipsative measures are used in psychometric tests allowing for respondents to choose from two or more options. However, these options are ideally given in as many statements. The disadvantage of questions of "quadri-modal" approach is that although they are de facto ipsative measures, they have been used in Likert scales. In such case, as Rudmin (2003) and van de Vijver and others (1999) have observed, the questions do not have null intercorrelations that are a prerequisite for conducting factor analysis.

7. Andhra Pradesh has since split to Andhra Pradesh and Telengana. This study refers to Telugu people based on their linguistic identity, belonging to undivided Andhra Pradesh.

Chapter Five

Methods

The study used mixed methods, wherein questionnaire survey and in-depth interviews provided quantitative and qualitative data respectively. The methods are informed by the circular model discussed in chapter 3, and the bi-dimensional acculturation model (Berry, 1997) discussed in chapter 4. In this chapter we describe the sample, data collection procedure, scales used in the quantitative survey, interview guide, and data analysis procedure.

SAMPLE

The sample that is the core of the work reported herein included 462 male migrant workers from Bangladesh and India. These men had been in Singapore for varying amounts of time. Only workers in semi-skilled and unskilled positions, employed in the industrial sectors of construction, oil refining, shipping, and other factory/plant related jobs were included in the study. Workers in these categories were given work permits, which represented the unskilled category, or "S pass," representing the semi-skilled category. The migrants were from four regions in the two countries, each with its own language and ethnic culture: Malayali, Tamil, and Telugu (from the south Indian states of Kerala, Tamil Nadu, and Andhra Pradesh, respectively) and Bengali (from India's northeastern region as well as from Bangladesh). We attempted to find a demographic breakdown of these workers based on their visa status from the Singapore's Ministry of Manpower. This was, however, not possible.

Among the 1.4 million foreign workers employed in Singapore, 0.3 to 0.5 million constitute the sample frame of workers within the low-skilled and semi-skilled categories, excluding maids. Earlier research in acculturation has used a wide variety of sample sizes irrespective of the total population under study: Kim (1977, 1978) studied Korean migrants in Chicago, US (a group consisting of approximately 20,000 persons) by analyzing 285 questionnaire responses; Hao & Zhu (2005) studied the media usage and adaptation patterns of Chinese immigrants in Singapore, who formed a subset of the broader 3 million people of Chinese ethnicity in the country, using a sample of 667 respondents.

DATA COLLECTION

Questionnaire and interview data were collected between March 2011 and June 2013. The questionnaire was translated to four languages, Bangla, Tamil, Telugu, and Malayalam; the initial translations and the back translations of the questionnaire were conducted by two different people who were proficient in both English and the respective language (see appendix C for the questionnaire in English). Among transient workers from South Asia, Bangla, Tamil, Telugu, and Malayali formed the largest groups. However, there is no disaggregate data to prove this. But based on authors' own perception during field work, and from the information gathered from respondents, this is largely true. Indians belonging to the Punjabi and Hindi speaking areas (north and northwest regions) were screened out, as they were too few in number to form a group of comparable size with the others. The questionnaire was administered by the first author personally. It was typically completed in 20 minutes, although some respondents took longer time to complete it, due mainly to their limited level of literacy and non-familiarity with survey questionnaires. The respondents were recruited through a mixture of purposive and snowball sampling methods, similar to other studies on mobile communication (e.g., Fortunati, Manganelli, Law, & Yang, 2008). Workers were contacted near their dormitories or shared homes in the evenings and weekends as they returned from work. After a respondent completed the questionnaire, the first author would ask him to refer his friends or acquaintances. At least 30% of respondents in the sample came through snowball sampling. Respondents who were found forthcoming in their response to the questionnaire were invited to participate in the in-depth interviews.

There is a chance that the respondents were sensitized during the data collection process, as both the survey and interview data were collected mostly from the same respondents. However, the particular situation of the research mandated that interviews be successfully initiated only after the respondent got an idea of the topic by filling out the questionnaire. In other words, sensitization was needed to make the respondents reflect on their experiences in the presence of an outsider—namely the first author. Moreover, the objectives of the questionnaire and the interviews were different: Whereas the questionnaire was intended to get a quantifiable measure of migrants' mobile phone use, level of adaptation, and acculturation attitude, the qualitative interviews probed further into descriptive assessments of their life experiences.

The minimum age for participating in the study was 21, determined by the University guidelines. A token amount of 5 SGD (1 USD = 1.36 SGD approximately) was paid to each questionnaire respondent and 10 SGD to each interview respondent. Funds for the research were granted jointly by an internet research group in Singapore and the University where the authors are based.

Most of the population under study resided in commercially run, purpose-built dormitories (MoM, 2017b). However, some resided in converted industrial premises or quarters on construction sites, which were temporary and makeshift. Due to these difficulties, a random sampling frame used in previous acculturation research (e.g., Hao & Zhu, 2005; Kim, 1977, 1978) was not feasible in the context of this study. Given the sampling approach and the nature of the social group, it is correct to say that it is not statistically representative. That said, it is prohibitively difficult to develop a statistically representative sample of such a diverse and ill-enumerated social group.

The actual response rate was not calculable for both questionnaire survey and interviews. However, broadly, one in five migrant workers approached by the first author agreed to fill in the questionnaire and one out of three workers agreed to participate in interviews. While disaggregate demographic data on temporary migrant workers in Singapore was not available, news reports and studies (Piper, 2005; Rahman & Lian, 2005) suggest higher proportions of Bangladeshi and Tamil workers. In addition to the two ethnicities, it is likely that there was an over-representation of Telugu in the sample (Adapa, 2002).

The respondents were approached after being identified as Indian/Bangladeshi. They were then asked about their sub-ethnic identities. If the workers' sub-ethnicities were one in the four under study (i.e., Malayali, Tamil, Telugu, or Bengali), they were briefed about the study and were allowed to take the survey after giving their consent. Due to some underlying individual and group characteristics like attitude toward outsiders, amiability, feeling of confidence, and the perceived value of the study incentive, some groups were more willing to participate in the study than others: in order of willingness, they were Bengali, Tamil, Malayali, and Telugu. Since the first author had set a maximum limit of eight questionnaire responses for each day of field visits, some eligible and willing respondents from the Bengali and Tamil groups had to be turned away, and thus more time was spent on less-willing and thinly numbered groups of Malayali and Telugu.

Initially 530 migrant workers participated in the study—519 filled in questionnaires and 75 participated in the interviews (with 64 interview respondents overlapping with the questionnaire group and 11 respondents for interview alone). However the sample was reduced to 464 after detecting missing data problems in some filled-in questionnaires.

The problem of missing data was handled using a combination of "deletion" and "simple imputation methods." The instruments drew largely from the Mutual Intercultural Relations in Plural Societies (MIRIPS) questionnaire,[1] which is part of an international collaboration project investigating feelings of cultural security, ethnocentrism, hierarchy, and reciprocity in multiple societies. When this questionnaire was used, there was a systematic exclusion of six questions among 79 respondents. These cases (observations) were omitted from the data set before statistical analyses, bringing down the sample size for quantitative analysis to 440.[2] The respondents who formed the intersection set of questionnaire and interview participants dropped from 64 to 51 after data cleanup (i.e., of the 79 observations omitted from the dataset, 13 were that of interview respondents).

QUANTITATIVE MEASURES

Acculturation Orientations

The survey was designed so as to examine *assimilation, integration, separation, marginalization, cultural identity,* and *multiculturalism in workplace* among the population studied. The five-point Likert scale questions (from 1 "strongly disagree" to 5 "strongly agree") were adapted for each of the ethnic groups. New questions were introduced to the original MIRIPS questionnaire items to overcome their ipsative[3] character, an issue raised by critics of the bi-dimensional model (Hicks, 1970; Rudmin, 2003). For example, a typical question on acculturation orientation has two dimensions that are to be traced on a one-dimensional Likert scale: "I feel that [ethnic group] should maintain our own cultural traditions but also adopt those of Singaporeans'." To overcome this problem, a few one-dimensional questions, enquiring either only about respondents' affinity to host culture, or to their affinity to home culture, were included in the questionnaire. The new questions include, "I have at least five friends from my own community in Singapore," "I have at least five friends from other dominant communities in Singapore," "I think I have become a part of Singaporean society," and "I will be happy to adopt Singaporean culture."

A factor analysis with varimax rotation and unrestricted number of factors was used to determine which items from the questionnaire best captured the acculturation types for this sample population, in addition to evaluating the cultural identity and multiculturalism of Indian migrants. Thirty-eight items were input for analysis. Any item with factor loading less than 0.4 was omitted. Although 10 factors emerged from the factor analysis, the items representing the four acculturation orientations were not meaningfully grouped around factors. Low Cronbach's alpha also made these scales unreliable (see table 3 in appendix A).

The low reliability score for the MIRIPS questions on acculturation orientations has also been reported in other studies. Studies have criticized the operationalization of Berry's (2005) bi-directional, four-typology framework. According to Rudmin (2003), "Questions about the fourfold paradigm have begun to arise, and it is increasingly under criticism" (p. 4). Citing a factor analytic study conducted by van de Vijver, Helms-Lorenz, and Feltzer (1999), Rudmin argued that the four acculturation scales in concert measured only one dimension, not two or four. It suggests that the original scale for acculturation orientations failed to capture the phenomenon in the population of the present study (see appendix E for more details on factor analysis).

Factors 1 and 3 related to strength of cultural identity and multiculturalism in workplace, respectively. Both factors had eigenvalue above 2.6. They were finally used in the statistical modeling. A description of these is given below.

Cultural Identity

Cultural identity denotes one's strength of loyalty towards one's culture and heritage. This is generally assessed based on respondents' level of agreement towards direct statements like "I think of myself as [ethnic]," "I feel that I am part of [ethnic] culture," "I am proud of being [ethnic]," and "I am happy to be [ethnic]." Different variations of these questions have been used by researchers to suit the specific requirements. For example, Ward and Kennedy (1994) developed an Acculturation Index, which had 21 items to assess respondents' identification to culture of origin and culture of contact, using questions on lifestyle, recreational activities, and perceptions like "Are your experiences and behaviors similar to people from your country of origin?" In the present study, the questions assessed migrants' loyalty towards their home culture in general.

Multiculturalism in Workplace

Multiculturalism in workplace assessed migrant workers' level of acceptance of other cultures in the organizational space. Respondents' level of agreement with direct statements like "It is a pleasure to work with people of different ethnicities" and "It is best for an organization if all workers forget their different ethnic and cultural differences and work in unison" provide a quantitative measure for this factor. Studies have used similar items in other cultural contexts. For example, Dandy and Pe-Pua (2010) used a 10-item Multicultural Ideology and Social Equality Beliefs scale to test Australian host society's multicultural character, which included questions like "Australians should recognize that cultural and racial diversity is a fundamental characteristic of Australian society" (p. 37). Van de Vijver, Breugelmans, and Schalk-Soekar (2008) used a Multiculturalism Ideology Scale, which provided "a short and reliable measure of majority attitudes toward multiculturalism" (p. 96).

Combined, "cultural identity" and "multiculturalism in workplace" defined the acculturation of migrant workers in this study. The following section describes the measures used for outcome variables, which are "adaptation to social life" and "organizational commitment."

Adaptation Outcomes: Social Life and Work Domains

Adaptation outcomes were measured in the domains of social life and work. Three constructs used to measure the adaptation outcome in social life are—*proficiency in host society language*, *adaptation to host country rules,* and *adaptation to host society's custom and lifestyle.* Similarly, in the work domain, migrant workers' organizational commitment was used as the index of adaptation outcome.

Adaptation to Social Life

The items to measure adaptation outcomes in social domain were drawn from the sub-scale "socio-cultural competence" in the MIRIPS questionnaire (Berry et al., 1989; Berry, 2010) and were supplemented with context-specific questions. Items like "How difficult it is to cope with the pace of life in the new country" have been split and changed to "How comfortable are you with the metro life" and "How comfortable are you with the city's night life." MIRIPS question "How difficult it is to follow the rules and regulations in the new country" has been split and changed to "How comfortable are you obeying the traffic regulations in the city as a pedestrian," "How comfortable are you with obeying the drug law," and "How comfortable are you with obeying the anti-littering law." The modifications to the scale pertained to specific social and governance conditions of Singapore.

Organizational Commitment

Organizational commitment was measured using a five-point Likert scale.[4] O'Reilly and Chatman (1986) conceptualize organizational commitment on the basis of three constructs: compliance, identification, and internalization. Compliance, the behavior aimed at gaining specific rewards, is measured using items like "Unless I'm rewarded for it in some way, I see no reason to expend extra effort on behalf of this organization" and "In my job here, I sometimes have to act in ways that are not completely consistent with my true values" (Kelman, 1958; O'Reilly & Chatman, 1986).

To measure identification with the organization, items like "I talk up the [organization] to my friends as a great organization to work for" and "I feel a sense of pride in working for this organization" were used (Buchanan, 1974; O'Reilly, Chatman, & Caldwell, 1991). The identification measure, when used separately, had a Cronbach's alpha of .862 in Buchanan (1974, p. 539). Internalization measure was created by O'Reilly and Chatman (1986) and includes items like "I find that my values and the organization's values are very similar" and "Since joining this organization, my personal values and those of the organization have become more similar." The scale with all constructs combined has a Cronbach's alpha of .94 in Buchanan (1974, p. 539). Mowday, Steers, and Porter (1979) administered the Organizational Commitment Questionnaire to employee populations ranging from hospital employees, auto company employees, and scientists and engineers and observed that the reliability alpha fell within the range of .82 and .93 (pp. 230–232). O'Reilly and Chatman (1986) showed that the three dimensions are orthogonal (p. 494).

The 19-item measure used in this study was derived from the previous studies with an added question "At times I wonder how I have begun to feel more committed towards my work after coming to Singapore," which reflected migrants' situation of a shift from their home country to Singapore (refer to table 5 in appendix A for factor analysis results and reliability scores).

Volume of Mobile Phone Communication

Migrant workers prefer to use cheap one-on-one communication with home and host cultures (see, for example, Lin & Sun, 2011; Thomas & Lim, 2011). Use of SMS is characteristic among female domestic workers who use their mobile phones during work breaks. By way of contrast, there is no employer control on male migrant workers with regard to the use of mobile phone, other than when they are working in dangerous work sites.

In this study we chose to focus on mobile voice calls when examining phone use. We did this for several reasons. In the Singaporean context, voice calling is equally cheap and has advantages when compared with SMS, such as faster communication and less room for miscommunication between migrants who have different linguistic backgrounds. Voice calling can be evaluated along two dimensions, that is, frequency and duration of calling, compared to SMS, which can only be assessed by way of average count per day or week. Another reason to choose voice calling as the quantitative measure pertains to the real-time interaction possibility it affords the user. Unlike SMS, which is asynchronous, synchronous voice conversations can simulate face-to-face interactions more closely than text messages. There was also the issue of literacy. As a number of respondents was less literate, texting was an uncommon practice followed by this population. It is for these reasons that the measure of mobile voice calls was used.

Duration and frequency of calls were used to describe one-on-one voice communication. The data was collected based on self-report items in the questionnaire. Duration and frequency of mobile phone calling, in addition to texting, were key indices of amount of mobile phone usage before the advent of smartphones among the migrant worker population during the last few years. Further, duration and frequency measurements are in line with the "amount of exposure" concept widely followed by media effect studies, cultivation analysis, and uses and gratifications approach (Shrum, 2002; Signorielli & Morgan, 1990). As migrants use mobile phones to fulfill informational needs, for networking, to strengthen intimate relationships, and for entertainment, both duration and frequency dimensions provide a comprehensive measurement of migrants' one-on-one communication. Duration and frequency of calling have also been used by studies that attempted to find out the level of involvement of mobile users (Ciunova-Shuleska et al., 2012; Haverila, 2011). In the present study, frequency and duration have been measured along the axis of cultural referencing; that is, calls that are likely to make home culture salient and those that make host culture salient. All the measures related to mobile communication were developed by the first author,[5] since there is little prior research combining cultural orientation and mobile phone use. Indeed, this is an original contribution of this work.

Mobile Phone Calling to Other Cultures

Six items were included in the scale, reflecting both frequency and duration dimensions with regards to calls to interlocutors from the host culture. Frequency items included, "How often do you call your friends from other communities in Singapore?" and "How often do you call your fellow-workmen, at or below your level." Responses ranged from "Virtually never" (0), "Once in a week (2), to "At least once in a day" (5). Duration items included, "For how long do you talk to your friends in other communities in Singapore?" and "For how long do you talk to your fellow-workmen, at or below your level?" Responses ranged from "Not applicable" (0), "Approximately 5 to 10 mts" (2), to "More than an hour" (5). The measure was internally reliable (Cronbach's alpha: .73).

Mobile Phone Calling to Home Culture

Eight items were included in the scale, reflecting both frequency and duration of mobile use. Frequency items included, "How often do you call your family back home?" and "How often do you call your friends from your own community in Singapore?" Responses ranged from "Virtually never" (0), "Once in a week (2), to "At least once in a day" (5). Duration items included, "For how long do you talk to your friends back home over phone at a time?" and "For how long do you talk to your friends in other countries over phone at a time?" Responses ranged from "Not applicable" (0), "Approximately 5 to 10 mts" (2), to "More than an hour" (5). The measure was internally reliable (Cronbach's alpha: .66).

QUALITATIVE INSTRUMENT

The qualitative interview followed the narrative research techniques designed to elicit the verbal recollections of the interviewees (Creswell, 2007; Mantovani, 2012), especially those related to their acculturation experience. The quantifiable findings from the questionnaire provided insight into the volume of mobile voice calls (e.g., frequency and duration of calls) and acculturation. The qualitative interview data supplemented the quantitative data drawing methodological knowledge from a range of research traditions from narrative methods to phenomenology.

The interview followed the tradition of narrative research in acculturation (Arasaratnam, 2008; Ryan et al., 2008; Skuza, 2007). The qualitative analysis allowed for an in-depth exploration of the acculturation-communication experience. The qualitative interviews allowed the migrants to describe their lives in the host country. It gave them the chance to describe whether they found themselves in collaboration or in confrontation with members of other cultures.

The languages used to conduct the interview include English, Hindi, and Malayalam. The first author had working knowledge of all three languages. Interviews lasted for an average of 20 to 30 minutes. The first part of the interview included questions on usage history (e.g., "How long have you been using the mobile phone?" "How has your communication over mobile phone to family, friends, and colleagues changed over time, especially during the migration phase?"), and their mobile skills (e.g., "What are the new skills you acquired over time to interact with newer and more sophisticated affordances of the phone?").

The second part of the interview focused on usage and handling of the mobile phone. Questions asked under this section included, "How has the mobile phone helped you maintain relationship with family and friends?" "How has the mobile phone helped you manage your life in the host society?" and "How do you use the mobile phone for pastime and distraction?"

The third portion of the interview explored the issues of prestige and social identity as they related to mobile telephony. In this portion, questions included, "Have you come across situations when you have been identified by others with a particular pattern of your mobile phone usage?" and "Has your mobile device helped increase your prestige among peers?"

In the fourth section of the interview protocol, there were questions regarding meta-communication (e.g., "How do you communicate with others about your experience of using the mobile phone?" and "How have you come to know about mobile services in the host country?"). Finally, the fifth part of the interview explored the acculturation orientations of respondents. Questions in this category included, among other things, "To what extent do you think you have adapted yourself to the culture of Singapore?" "How does your relationship with the host communities in Singapore affect your identity as a Bangladeshi/Malayali/Tamil/Telugu and/or Indian?" "How do you maintain a bonding with your family back home with the help of mobile phone?" and "To what extent have you adapted to the multicultural workforce of your organization?" (see appendix D for the detailed interview guide).

The interviews were recorded and translated to English by the first author. They were then transcribed and coded on the basis of theorized categories—some pre-existing and some emergent. The information obtained was interpreted and categorized according to the themes in the context of a theory or conceptual framework. The theoretical foundation for the coding drew on various traditions including social constructionism, symbolic interactionism, phenomenology, and interpretive research, especially in the data analysis phase. For instance, social constructivism posits that text, technology, phenomenon, and events are not directly accessible. Rather we construct meanings from observations and other sensory data, wherein the meaning making process is influenced by sociocultural norms (MacKenzie & Wajcman, 1999). Building on this argument, the research demarcated two instances of meaning construction: (i) the way in which respondents imagined and developed meanings of their acculturation experience and technology use, and (ii) the way in which the authors interpreted the data after deriving meaning from the interview transcripts. Qualitative data in itself does not provide a quantifiable picture of the acculturation-appropriation situation. But when read along with the quantitative survey data, qualitative interviews provide a deeper insight into the phenomenon.

Symbolic interactionism emphasizes the meanings derived by individuals in representing themselves, while acting and responding to the demands of everyday life (Denzin, 1992; Hall, 1980). The inquiry guided by the symbolic cycle of appropriation in the circular model contributes to the approach of symbolic interactionism methodologically and epistemologically. Phenomenology addresses the lived experiences of individuals without reducing them to a priori psychological categories (Skuza, 2007). The study followed this approach while exploring the lived experiences of migrants as they interacted with members of the host and home cultures on a regular basis in the mediated spaces allowed by the mobile phone, as also in the unmediated spaces in the social and work domains. Finally, following an interpretive approach, the strict contours of Berry's typology were disregarded to allow for a more fluid categorization of the acculturating individuals.

During the thematic analyses, patterns of topics or issues that emerged with regard to the respondents' use of mobile phone were noted (Creswell, 2007). These patterns revealed how the mobile phone usage patterns related to the circular model and how mobile phone appropriation is linked to migrant acculturation at different levels.

The study follows established conventions in data analysis and the reporting of the qualitative findings. In research that draws from phenomenology and follows an interpretive approach, the convention to report findings is to consider any meaningful phrase or sentence that reflects a theme, or tells a story that informs the research. For example, de Souza e Silva and others (2011) analyzed the qualitative data pertaining to mobile phone appropriation using quotes from respondents. Skuza (2007) analyzed data by "(a) reading the data, (b) dividing the data into parts, (c) organizing and transforming the data into disciplinary and language, (d) expressing the structure of the phenomenon" (p. 450). Leonardi (2003) analyzed interview data by looking for themes, patterns, and positions built on cultural orientation. "Particular attention was given both to comments or stories that made claims about the purpose of a certain technology as well as its benefits and hindrances, and to the way in which participants framed their knowledge about the specific communication technologies" (p. 166). In this research, since the qualitative data were used both to triangulate the results of statistical analyses and to derive acculturation and appropriation types, it was important to provide detailed narrative descriptions for the relationships and patterns.

Recent Changes in Mobile Communication Landscape

We acknowledge the limitations of our qualitative instrument, which was developed before the widespread adoption of smartphones among the migrant workers. In view of this change, we did four additional interviews with respondents who have started using the smartphones since 2013. Major changes were reported in the appropriation part, wherein different strategies were adopted by the respondents for learning the multiple affordances of their new phone. We discuss these experiences at relevant places in chapters 6 and 7.

DATA ANALYSES

Quantitative and qualitative data were analyzed based on the theoretical frameworks discussed in chapters 3 and 4. With the quantitative data, following data cleanup and imputation of "series mean" to missing data points, regression diagnosis tests were run to check for normality, heteroscedasticity, and multicollinearity between variables. A detailed description of the diagnostic tests has been included in the following section. Regression tests were run on two dependent variables to test hypotheses 1 and 2. The two steps—regression diagnosis and regression tests—constituted the quantitative data analysis part of the study. Qualitative data analysis involved triangulation of the regression results with the qualitative data. The qualitative data was also analyzed to identify non-essential types of mobile phone appropriation and acculturation and to identify the links between the two phenomena.

Regression Diagnosis

Multicollinearity between independent variables cultural identity, multiculturalism in workplace, mobile communication to home, and host cultures, as also between the demographic variables, was within the acceptable limit (Variance Inflation Factor being less than 1.6 in all cases, except in the case of interaction terms). Descriptive statistics showed that both the mobile communication variables—amount of calling to home culture and amount of calling to dominant host cultures—followed a normal distribution. However, the two acculturation variables—cultural identity and multiculturalism in workplace—showed skewness towards higher values: For cultural identity, scores between 4.5 and 5 constituted 60.2% of the sample, whereas for multiculturalism, scores between 4.5 and 5 constituted 46.4% of the sample. As the skewness was not caused by outliers, the problem of non-normality of the two variables was considered to be a non-issue in the regression analysis.

Regression Tests

The questionnaire data were used to test the hypotheses on the effect of acculturation factors and mobile communication variables on adaptation outcomes (see figure 4 in appendix B for beta coefficients on the model). Hierarchical regression analysis was used to identify the relationship between the variables. Demographics was the first block in the regression equation. Age, education, duration of residence, and income were considered as one block in alignment with existing practice in acculturation research (Kim, 1977; Sunoo, Trotter, & Aames, 1980). The second block was cultural identity and multiculturalism in workplace, as both these variables tended to evaluate the acculturation attitude. The third block was mobile phone calling to home and host cultures, which are both related to the influence of mobile phone communication. In addition, a fourth block tested interaction effects between cultural identity and calling to host and home cultures and multiculturalism in workplace and calling to home and host cultures. Scores for all the six independent variables were centered before calculating the interaction terms. Two tests were run for two dependent variables of the study: adaptation to social life and Organizational commitment.

Qualitative Data Analyses

Qualitative data were analyzed to triangulate the results of the regression tests. As noted, a meaningful phrase, sentence, or paragraph was considered a unit of analysis (de Souza e Silva et al., 2011; Skuza, 2007). Themes and usage patterns were allowed to emerge as the data analysis progressed (Lindlof & Taylor, 2011). Patterns of mobile phone usage and appropriation among migrants were identified from the interview data, as also patterns of acculturation (for a detailed list of themes that were pre-determined, as well as categories that emerged at the time of data analysis, see table 10 in appendix A). These patterns were then used to identify user types and acculturation types among the respondents.

A critical discourse analysis perspective was followed while analyzing the data, wherein the hierarchical structures that constitute discourse and practice related to mobile phone use among migrants and their acculturation were brought to the fore (Chouliaraki & Fairclough, 1999; Fairclough, 2003; Foucault, 1981, 1991; Hammersley, 1997). Fairclough (2003) provides a manifesto for critical discourse analysis highlighting the following questions:

> How do [existing societies] deny people the possibilities and resources for rich and fulfilling lives . . . What possibilities are there for social change which would reduce these problems and enhance the quality of the lives of human beings? (p. 202)

Of particular importance in critical discourse analysis are social structures and social practices. Social structures help determine the way in which each member of a society acts and thinks, to the extent that these structures are a kind of external imposition on the members. Social practices are defined as a "relatively stabilized form of social activity" (p. 205). Within social practices, Fairclough (2003) identifies different elements that are interconnected (e.g., subjects and their social relations, values, and discourses). It is the relationship between social structures and social practices that is attempted in critical discourse analysis. According to Toolan (1997), discourses are "symbolic capital, exploited in ways that benefit some and disadvantage others" (p. 89). The objective of critical discourse analysis is to reveal the ways in which this symbolic capital is exploited in different communicative events so as to change the order of discourses itself.

For Foucault (1981, 1991), "discourse" has a broader meaning and implication, which implies neither a multiplicity of intended meanings nor muted words and expressions. Instead, "discourse is constituted by the difference between what one could say correctly at one period (under the rules of grammar and logic) and what is actually said" (Foucault, 1991, p. 63). Discourse pervades human activities that fall under three practical systems characterized by three relational spheres: "relations of control over things, relations of action upon others, [and] relations with oneself" (Foucault, 1997, p. 318). An investigation into these three areas forms the primary objective of critical discourse analysis, which ultimately leads to identifying the discourses that sustain the hegemonic structures embedded in social practices. The present research also follows the principles of critical discourse analysis, as it highlights the hierarchical structures that the migrants have to negotiate as they acculturate to the host society and use mobile phone to adapt to work and social life.

SUMMARY

This chapter presented the mixed methods approach followed by the study. It described the population, sample, the rationale behind snowball and purposive sampling methods and the data collection techniques adopted by the study. The chapter explained how the scales of cultural identity, multiculturalism in workplace, mobile phone calling to home and host cultures (independent variables), adaptation to social life, and organizational commitment (dependent variables) were created based on factor analysis and reliability test results. The chapter also described the qualitative instrument used in the study, especially the interview guide used to elicit data on acculturation and mobile phone appropriation. It also elaborated on the quantitative and qualitative data analyses methods followed by the study. Finally we have provided detailed demographic information for the sample, in terms of the respondents' age, educational level, duration of residence in the host country, and income categories. The following chapter presents the results of the analyses, to help answer the research questions.

NOTES

1. MIRIPS is led by John Berry. For details, see http://www.victoria.ac.nz/cacr/research/mirips

2. This method of listwise deletion (also called "complete case analysis") is recommended because of its simplicity, but also due to the advantage that it ensures equal sample size across analyses. This is different from "pairwise deletion" wherein the other variables of a case, except the one where the missing data was observed, would be retained. Pairwise deletion would lead to variable sample size during analyses. The questions that were systematically omitted by the respondents were: "It is not important for me to be fluent either in [ethnic language] or language used in Singapore," "I prefer social activities which involve Singaporeans only," "I feel that [ethnic group] should maintain our own cultural traditions but also adopt those of Singaporean," "I prefer to have only Singaporean local friends," "I prefer social activities which involve both Singaporeans and [ethnic] members," and "I prefer to have both [ethnic] and friends from dominant communities in Singapore." One possibility for systematic non-response is that the difficulty in choosing between the two cultural dimensions in the same question would have motivated the respondants to skip the question.

After omitting the 79 observations, the remaining instances of missing value were rare; no item had more than 3% of missing observations. Hence, these missing data points were imputed with "series mean," a single imputation method widely used in social science research. This method was adopted because the imputation would not end up changing the mean for sample for that particular variable. It also supports a complete analysis of the data. However, "series mean" imputation has the disadvantage that it reduces variability within the variable. Other methods include "hot-deck imputation" or "last observation carry forward," "cold-deck imputation" and "regression imputation," and model-based techniques like "maximum likelihood" and "multiple imputation." These techniques had limited scope, and were more traditional. Hence, considering the tradeoffs involved in choosing an imputation technique, the "series mean" imputation technique was finally chosen. The mean of all items in a scale was calculated after imputing the series mean to each item individually.

3. Ipsative measures ideally should provide two answer choices for the respondent related to each dimension measured in the question. The respondents then choose the best option that suits them. Instead, the failure of MIRIPS questions was to provide two dimensions in a Likert-scale. This is different from the problem of double-barreled questions which are considered an informal fallacy. Usually a compound question, the problem with double-barreled question can be solved just by splitting the question into two sentences. Whereas, in the case of MIRIPS questions,

the problem is more rooted in the constructs of the "quadri-modal" approach—*assimilation*, *integration*, *marginalization*, and *separation*.

4. From 1 "totally disagree" to 5 "totally agree."

5. The first author was guided by his supervisor in the doctoral programme. Any reference, future utilization, or adaptation of this scale may give proper credit to Dr. Arul Chib, along with the authors of this volume.

Chapter Six

Mobile Phone Appropriation

In this chapter we examine the appropriation of mobile phones using interview data. We will examine the process through which the workers purchased phones and, as described by Bar, Pisani, and Weber (2007), made them "their own," integrated them into their everyday lives, and also made them into an element of their status identity. This was based on the circular model, and its different components such as "usage and handling," "prestige and social identity," and "usage and symbolic cycles" (See the circular model, figure 1, appendix B). Appropriation is the foundation on which other findings of the study can be based (e.g., on how the mobile phone is used to meet various ends including social adaptation, multiculturalism at workplace, and the development and retention of ties with home and host cultures).

In order to explore mobile phone appropriation we posed the question, "How do acculturation types relate to mobile phone appropriation types among migrants, when defined in non-essential and non-universal terms?" In this chapter we discuss the portion of the research question associated with the appropriation of the mobile phone by following a framework that goes beyond the functional uses of the device. Specifically, we were interested to examine the issues of "usage and handling" as well as the issues of "prestige and social identity" that arise from owning a mobile phone. Based on this we suggest a taxonomy of use types (see table 10 in appendix A for the list of mobile phone appropriation themes that emerged with data analysis). However, to understand the appropriation phenomenon in depth, an overview of different considerations and strategies of migrant workers with respect to their mobile phone use is needed. We begin with a description of conditions that limit mobile phone use among migrants, before discussing the various use patterns.

ISSUES THAT LIMIT AND CONSTRAIN MOBILE PHONE USE

Among a number of factors that posed a limit on migrants' use of the phone, a few included: (i) cost of calls; (ii) safekeeping of the phone; and (iii) workplace rules. Cost related factors involved a primary consideration by the users to maintain a balance with monthly earnings and their mobile phone use. Concerns related to safekeeping of the phone were more pertinent to that era of smartphones wherein the cost of these advanced devices were considerably higher than that of feature phones. As a result, users took extra care to keep their smartphones safe, for which they had to strategize their use and non-use. Workplace rules decided who was allowed to use the phone while at work and who was not.

Cost Related Issues

The material showed that the respondents balanced intensity and frequency of use of mobile phone with cost considerations. The interviewees, to a large extent, used low-cost pre-paid cards for international calling.[1] They actively aligned the expense for day-to-day mobile phone use with their monthly earnings. The major budget post for telecommunication was for in-country and international calls.[2] On an average a worker who earned about SGD 700 (approximately US$ 515) a month, including overtime, spent SGD 30 to 50 for mobile phone calling. This range varied in special cases, for example, in the case of better-educated workers who subscribed to broadband internet, or in the case of a worker who talked at length with his newly wedded bride or fiancée.

Those who had spent several years in Singapore understood the reduced tariff of mobile calling and other communication methods from a historical perspective. Khurram, 32, a Bangladeshi construction supervisor explained:

> When I come first time, that time was so expensive, lah. Later then now cheap. Right now also cheap. That's sixteen years back. I come the first time don't have [I didn't have mobile phone when I came here the first time]. Then I got pager. For landline, I inserted coin. I bought a handset in 1998. I used the handset for more than ten years. After 2005, the international call rates became less.

As a consequence, the workers reported having the luxury of more time for international calls, which was mostly utilized to talk to family:

> The call rate was higher when I arrived here first in 2005. The amount of talking time was less. Now, the average amount of time we talk to people back home has increased. When there is more time to talk, I can talk more with family.

> We can talk more family matters. Money can be saved. (Ganesh, 30, Malayali-Indian worker[3])

Even though international call charges dropped since 2005, cost considerations continued to determine the contours of the respondents' use of their mobile phones, especially when they surfed the net. For example, some respondents recalled how they avoided browsing the mobile internet, or sometimes limited its use for fear of making a dent into their phone credit, as seen in the comments of Telugu-Indian worker Balakrishna:

> If I use internet on mobile, it costs more . . . If I download something, the local balance is used up very fast. So I go to the [internet] café when I feel like. (Balakrishna, 25)

Cost concerns were shared by migrants' families too. They adopted strategies that made the migrant member pay for the international calls. This was acceptable to both parties considering that the calls from both Bangladesh and India to Singapore cost more in absolute and relative terms. Hence, as an earning member, the migrant worker was willing to cover this expense.

Internationalized Missed Calls

As noted, there had been a reduction in the cost of calling over the years for the migrants. This reduction internationalized the practice of "missed calls," widely reported as a developing country phenomenon in literature (Donner, 2007; Sey, 2011). This meant that an eventual interlocutor from the migrants' home country would place a "missed call" at the migrants' phone, to alert them that they would like to talk to the latter. The "missed caller" then hung up before the call was answered. When the worker in Singapore saw that the call was from someone he knew in his home country, he would assume the cost of the interaction by calling back. The person in Singapore would then use an international calling card to call back, thus saving money. This system was commonly used as a way to push the cost of the interaction to the person with the greater economic wherewithal.

However, the system of "missed call" also had its own limitations, mainly due to the workers' inability to attend to calls during work time. In such cases, the "missed call" acted as a message sent free of cost to the worker notifying him of the need to call back, at his convenience, but without fail:

> Work time my house any problem, eh . . . [they] call "missed call." After I see what time. So "I have work, I cannot answer." Makkan [lunch] time, I answer. Lunch after doing, eh [If it's after lunch], so, 4:50 I work finish. After I call. Coming time I call [I call when I return to my dormitory]. Important time, eh . . . many

> many call. Two time three time. Normal time one time call, ok. So, I understand what happened. So, two time three time call, eh, coming time I call. One time call, eh, I coming house I rest, I makkan after I call. So, two three missed call means something urgent. Not family members alone, but everyone knows, all my friends, everyone. (Salim, 26, Bangladeshi, Construction safety supervisor)

In addition to the implicit idea that the Singapore-based worker will return the call, there were also implicit meanings involved in the number of "missed calls" from a sender. These meanings were determined by the number of "missed calls," the interval between two "missed calls," and other codes agreed to between the two parties. Beyond these implicit meanings, "missed calls" also signified a broader change, as the medium itself became a message in a McLuhanian sense. For Chandran, this meant that compared to his stint in the Middle East, he had better opportunities to communicate with his family:

> I used to talk less to family when working in Dubai in 2007. The call rates where high . . . I was also earning less those days . . . I used to send letters. After coming here [Singapore], I don't write letters. I call home every day, sometimes even thrice a day. I may be talking for a short time, maybe for 3 minutes or 5 minutes . . . When my daughter returns from school, she "missed calls" me immediately. Even if I return late at night from work, I have to call home; otherwise there will be a barrage of "missed calls" from her. It is a pleasure being constantly in touch with family. (Chandran, 45, Malayali-Indian, General worker)

Rather than "missed call" representing a constraint, it constantly reminded Chandran of his enhanced ties with family. For the migrant families, "missed calls" were more than a cost saving technique. The "missed calls" conveyed their message of love and concern to the member abroad.

Concerns Regarding Safekeeping of Phones: Multiple Handsets

At least 20% of the workers were in possession of two mobile phone handsets: one for personal use and the other for work-related use. Indeed it was often the case that the interviewees saw having a phone as an implicit job requirement. Respondents differentiated between a basic handset (often a simple feature phone) meant for rough use at a work site and an advanced handset (often a smartphone or a touchscreen phone) for use during leisure time. The advanced (and often more expensive) mobile device was most often left in their rooms under safe conditions during the workday, only to be used in the evenings for accessing entertainment content (e.g., watching films or sporting events).

Bangladeshi worker Jaffer, 24, like many other respondents, had clearly differentiated between his touchscreen phone, which was meant to "listen to songs, play games, [and] sometimes use internet," and the feature phone that he carried to the worksite, always with an apprehension that "while working, the phone can drop." Although "rough use" of a feature phone suggested use in rough conditions like a construction worksite or a welding pit, it also suggested a cheap mobile device that can be used with less caution. Such a low-cost mobile phone had little emotional value for the users and was evaluated only on the basis of its material worth, demonstrating how the symbolic cycle was linked with the use cycle of the circular model. Selvan, 49, a Tamil-Indian who had been working in Singapore for 18 years, echoed these sentiments:

> I leave my iPhone in my room when I go for work. It's gifted by my brother-in-law, who is in US. I carry this basic phone. This is a new one, I bought it a year ago. I don't mind what happens to it . . . it can be changed any time.

While it may seem counterintuitive that having two phones was based on considerations of cost, the informants framed it in this way. They saw it as a way to avoid destroying the more expensive device.

Workplace Rules Limiting Phone Use

Workers also faced limitations at worksites on the use of phone. There were different dimensions to this prohibition. In the freest scenario, the workers were allowed to use mobile phone at all times during work. In fact, this helped them coordinate work with their team. We shall return to this discussion later in the chapter. At the other extreme, there is total prohibition of mobile phone use in potentially dangerous worksites (e.g., an oil plant which had live electric wires, or on the deck of a ship under repair). Use of phone in such prohibited areas would invite warning and fine. Workers would lock their phones away during work, often switched to "silent" mode, and return to check them during breaks. A long "mobile-free period" was preceded or followed by intense communication to family and friends.

In addition to these two cases, companies selectively imposed rules on mobile use, in order to ensure worker productivity. However, our respondents opined that this, in fact, had a negative effect. When workers would benefit from watching a video related to their work, a ban on mobile phone use negatively affected their performance. Victor, a 25-year-old Malayali-Indian welder, explained how access to multimedia during work helped improve his work performance:

> What we have in our course-related books is only basic information. Anything beyond that can be accessed by going to the Welding Society's website . . . It's really vast . . . Suppose we want to know what "welding crack" is or, what "welding discontinuity" is, while doing our work . . . But we cannot take our laptop to the worksite. So, isn't it amazing if we get information on the phone, on the spot?

Workers like Victor are different from Jaffer and Selvan who had no opportunity to gain from using smartphones at worksites. Victor also recalled how dictionary App helped workers like him to find meanings of English words as soon as they heard them. In the absence of smartphones, they had to scrawl down English words furtively on their palm, or take a chance with their memory, as they returned to dormitories and looked for their meanings in a print dictionary or on their laptops. Victor went one step ahead and found out that there are Apps that helped him learn "industrial terms," something that directly helped him in his job.

These benefits notwithstanding, mobile phone calling and multimedia use were banned by some companies. A few respondents defended the ban on mobile use, based on arguments that accessing multimedia content during work could lead to workers' inattentiveness and distraction, with catastrophic results. This muddied the issue of a ban on mobile use during work, especially when the workers were not convinced of the reasons behind the ban.

Deepan, a senior employee, supported the ban arguing that it was part of the safety regime of the organization:

> Some employees misused it. As a result, only a few senior people like us have been given the pass by the company to use mobile phone. . . . It's not because they spend long time over phone calls . . . it's not calling that is problem . . . today's camera phone and other features create a lot of problem . . . employees become increasingly inattentive towards work. That's why the company has decided thus. In my opinion, it is right to decide thus. People sitting on scaffolds and other high-rise stands . . . watching movies, listening to songs. Such acts will have serious adverse consequences. When there is an accident . . . say fire, one will not recognize the siren, if he's sitting and listening to a song. (Deepan, 40, Malayali-Indian)

While there were safety related justifications, it was also possible to detect in the material a subtle logic that the migrant workers were incapable of differentiating the good from the bad; the useful from the frivolous.[4] The restriction was also imposed in a kind of paternalistic manner, more often couched in words suggesting that the workers were being helped, protected, and looked after in a brotherly manner. Pranab, a 40-year-old Bangladeshi, worked as a supervisor in a construction site. Only he and a few senior employees were allowed to use a mobile phone at work sites. Pranab regularly "helped" workers in his team understand the negative consequences of violating the prohibition:

> Only for us we can use [the mobile phone]. Because I talk to my boss, project manager [. . .] Workers must stop there, cannot use [. . .] Any worker use the phone, fine 200 dollars. That's why we every morning briefing to safety: "Any workers don't use the phone." Fine any letter come, f**k me also, because every morning never briefing [. . .] We. That's why I want to inform the men: "Man never use [. . .] You come for work, don't use the [phone] [. . .] and waste money, because you work very hard."

Pranab's words suggested the intricate ways in which the mobile phone ban was imposed and enforced among workers in the company. The workers were constantly reminded of the economic downsides of flouting the rules. While the rules were internalized, the workforce was made more docile and the possibility of any kind of resistance was nullified. The interviewees noted that they had to tread a narrow line to avoid reprimands from the company for flouting rules on mobile phone use and uphold personal work ethics while accessing the multimedia content.

That said, the situation of the male migrants was somewhat different from that of the female domestic workers in Singapore. The male migrants, generally, had more autonomy in their use of mobile phone (Thomas & Lim, 2011). In the case of female domestic workers, employers frowned on use and in some cases even impounded their mobile phones. There was the sense that they would be distracted from their duties. In the case of male workers, the justification to limit use of mobile devices was in view of safety and security.

In sum, at the time of data collection between 2011 and 2013, many migrant workers were still using basic mobile phone functionalities like voice calling and text messaging. At that point, the diffusion of smartphones was at a nascent stage. The material showed that a migrant worker might spend up to 10% of his monthly income for regular mobile phone expenses. However, within this allocation, migrants were mindful of wasteful use. Concerns related to cost were later internationalized, when migrants' families back home increasingly used "missed calls" to connect with the migrant in Singapore. The section also looked at other limitations of mobile phone use like safekeeping of the smartphones and workplace rules restricting the use. The following sections go in-depth into the appropriation question, examining the phenomenon with the help of the circular model.

USAGE AND HANDLING

Analysis of the interview material revealed four goal-oriented mobile phone uses. These included uses aimed at: (1) *distraction/pastime,* (2) *maintenance of relationships,* (3) *management of everyday life,* and (4) *wielding control.* These were determined by respondents' skills and motivation to use the mobile phone, their economic conditions, as well as the restrictions posed by social and organizational norms.

Use of Mobile for Distraction/Pastime

For many respondents, embedded features of the mobile phone allowed for more opportunities for distraction with which to while away their free time. As discussed above, the entertainment opportunities available on mobile phones were a bone of contention between organizations and employees. But outside work, multimedia, social networking sites, and the internet in general provided the best unchecked opportunity for the migrants for entertainment. While some respondents actively used these functions, others, due to limited financial resources and lack of technical skills, lagged behind.

The interviewees reported using their mobile phone's multimedia functionality to fill in low-activity periods during journeys or in between work assignments. Mobile phone is used to fill time when there were no other activities to which they needed to attend. This could be short pauses during work or longer times when off work. Malayali-Indian worker Lawrence, 26, had a 15- to 20-minute commute to work every day, which he utilized by listening to vernacular songs downloaded to his mobile phone. He also noted that downloaded songs on a mobile phone came in handy for him when "I'm in a mood for sleeping."

Advanced pastime uses of the mobile internet also included reading vernacular newspapers. By accessing these newspapers via their mobile phones, respondents could make informed judgments about the social and political situation of their home countries, which helped them improve their standing in informal groups (Aricat, 2015):

> This option is called "news hunt" . . . especially for those coming from places like India, to receive the news from their home country in their native language. It updates news every hour. In my case, for the last two months, in view of my heightened activities on "news hunt," especially for checking the election results, local balance in my phone card has been used up quite fast. Many of my friends use this facility. We then have issues to discuss on a daily basis. (Kesavan, 35, Quality Controller, Malayali-Indian)

Accessing news on mobile phones was less widespread at the time of these interviews four to five years ago when data for this study was conducted. Thus Kesavan's experience as narrated above was not typical of majority of the workers interviewed at that time. However, our additional follow-up interviews showed that news Applications on smartphones had been diffusing among this group of migrants rapidly in the last few years. News was also accessed from other media channels including print newspapers and television. Irrespective of the media used to access news, the fact remained that it fueled discussions among the migrants, especially co-nationals and co-ethnics.

Not all the "entertainment" use was associated with downloaded or online multimedia content. The calling and texting functionality of the phones also helped the workers to while away leisure time. In some cases, the device was used for coordinating social interaction. In the words of Prakash, 25, a Tamil-Indian worker, "We have life outside work . . . for instance, tomorrow they [my friends] have arranged for a trip to Marina Bay. They called me and informed me." The social/leisure use of the device manifested itself in a plethora of ways, ranging from light banter between friends to coordination among migrants in view of an entertainment event in offline space. In another case, Siddharth, a 25-year-old Telugu-Indian worker, acknowledged that he had regular mobile phone communication with a Filipino woman, whom he had befriended a few months earlier. He noted that in addition to the social interaction, their discussions had substantially improved his English proficiency:

> After I came to Singapore my English improved. I use mobile phone to talk to my girlfriend most of the time. It helped improve my language also. In fact, I always had interest in learning English. Mobile phone helped me improve it, 90 percent.

Hence, distraction and pastime use of the mobile phone had extended effects on migrants: The device provided a cultural learning opportunity for the migrants (as seen in Siddharth's acquiring of English skills), and they helped the interviewees in their interactions with co-ethnics (as in the case of migrants using vernacular news websites or Apps to access news and discuss it among co-ethnic migrants). Underlying most of these uses is the enhanced opportunity to connect with others and to maintain relationships. This we examine in the following section.

Use of Mobile for Relationship Maintenance

A number of relationships were valued by the respondents, including those with their family and friends back home as well as friendly ties with co-ethnics and coworkers. As a result of inexpensive international calling rates, there had been a spike in the number of outgoing international calls (Vertovec, 2004). Our material showed that this has also included, where possible, newer forms of interaction like video chatting, social networking, and voice over internet calling, associated with the mobile internet diffusion. This newly found telephonic contact plays an important role in the lives of both the workers and their families back home. Indeed, as Donner (2009) noted, calls and "even non-calls[5] might matter, as reachability itself (social presence at a distance) can play a role in determining people's choices about how to conduct their day" (p. 95). In the material examined here, expressive interaction between the interlocutors helped maintain and strengthen relationships:

> A life without mobile phone!! . . . I would definitely find it difficult to survive without It . . . if I don't call home daily, they will feel sad . . . I also feel sad, if I don't call there everyday, I'll keep thinking about what's going on there. (Deepan)

The role of expressive interaction is discernible in Deepan's comments. While instrumental exchange of information between the callers was important, the expressive side was clearly central.

The material indicates that migrants maintain relationship with coworkers from different countries, which was carried out through subtle acts of caring. This is seen in the comments of 29-year-old Tamil-Indian worker Shekhar:

> Two days back my [Burmese] friend went to the hospital . . . some chest problem . . . today morning I called him. He was hospitalized for three days . . . now he's discharged. When I called, I asked him how he felt . . . whether he feels better now. (Shekhar)

Kathiravelu (2012) identified the importance of such care networks outside the institutional mechanisms among migrant workers in Dubai. She noted that they helped ameliorate the workers' marginalized existence. The comments of Shekhar indicate that the informal care networks are also to be found among migrant workers in Singapore.

In short, mobiles are used to ensure social presence at a distance and as a channel through which to interact with co-ethnics as well as migrants' close family and friends. Together, they help develop and strengthen friends' networks outside the institutional mechanisms. These networks have a decisive role in cultural adaptation, as when migrants manage their life with limited resources on a day to day basis.

Use of Mobile for Management of Everyday Life

Most respondents organized their daily life by using a wide variety of functions on their mobile phones, at times the only communication device available to them. Workers were transported to work sites early in the morning so as to avoid peak hour traffic. Because of the need to wake up early in the morning, they used the alarm function of the mobile phone. Similarly, emergency messages were passed on to the management using mobile phones. In addition, respondents noted that they often used their mobile phones to inform employers about sick leave. For example, Hussain, a 25-year-old Tamil-Indian, noted "If I am sick, I call my supervisor and tell him: *I cannot make it today, I cannot make it to work*. It's allowed to take leave by calling on phone." Informants also noted that they used their mobile phones to arrange logistics as they moved between work sites and company headquarters.

As is obvious, the policy of "bring your own device," was quite often the policy adopted by the employers (Stephens et al., 2017). Most respondents used their personal mobile phones to make work-related calls, for which they never sought reimbursement from the organization, nor was it offered. In some cases, however, lower status respondents were generally not allowed to use a mobile phone during work, as we noted earlier. If there arose a reason for them to make a call, they needed to approach their team leader, who would either have a walkie-talkie, or a mobile phone or both provided by the company. This is seen in the case of Afsal, 43, a Tamil-Indian. As a supervisor in refinery, Afsal had three mobile phones: a smartphone, a feature phone, and a walkie-talkie. The smartphone was to access internet and to call friends, the feature phone to call family, whereas the walkie-talkie, which also had mobile phone functionality, was specifically a work phone. Thus for him, managing everyday life included making communication choices from a rather complex ecosystem of devices.

In other, often white collar, work contexts, smartphones supposedly challenged the traditional demarcations of home and work, "freeing one from the cubicle but also allowing work to seep into every nook and cranny of personal life" (Agger, 2011, p. 120). The material examined here suggests that many respondents wanted to demarcate their personal life from work life. By following a variety of strategies, such as turning off their mobile phones after work, or not picking calls that they think might have some relation to work, they were able to demarcate the two. This was particularly the case for workers who were not in supervisory positions (Chib & Aricat, 2012).

In the following section we will examine how management of daily life using mobile phone also helped migrants achieve greater control in their life, although the device sometimes subjected them to the control of others.

Being Controlled and Controlling Via the Mobile

Control, in the context of the workers' mobile phone use, had a range of meanings. One form of control implied the extent to which employers, aided by mobile phones, controlled the migrant workers, determining their movements, deciding their work schedules, and micro-managing their activities (Wallis, 2013). Another form of control involved a kind of subtle and reciprocal control experienced by the migrant workers from their family, a kind of "emotional tax" on the migrant workers who ensured their social presence in family with the help of mobile phone. A third form of control implied the migrant workers' greater control over their own lives in the host country, facilitated by the possibilities opened up by mobile phone.

As noted above, employers' control of migrant workers in this study was not as widespread as in the case of domestic workers (cf. Lin & Sun, 2011; Thomas & Lim, 2011). The workers in this study largely made use of mobile phones to turn the demands of work in their favor. This can be seen in the comments of Shekhar, 29, a Tamil-Indian, who worked as a fitter and rigger signalman:

> I know people who often receive such calls [outside work]. . . . Some pipe leaking, steam open . . . they have to go back and attend to the problem. Sometimes they rectify the problems over mobile phone also. (Shekhar)

In such cases as described by Shekhar, mobile phone emerged as an effective means of managing the demands of their employers. Employees, particularly those who had supervisory positions, risked being contacted by the management outside work hours if something was amiss. As Shekhar's comments indicate, some workers could fix the problems remotely, thus turning a potential issue of control in their favor.

Another dimension of the control is when the migrant's family exerts soft control, restricting the worker's ability to make economic decisions for himself. This happens in the case of workers who have been in Singapore for a considerable amount of time, which provides an impression that the migrant is an economic resource that can be tapped into in times of need. Vineeth, 23, a Malayali-Indian worker, recalled the difficult times in the past when he could hardly find money for mobile phone top-up. But whenever he established contact with family after overcoming the economic obstacles, his family failed to be sympathetic towards his plight: "When I call home they say, *whatever it is, you stick to the job there for two years. . . . Look for another job.*" Similar cases of heightened expectations from family controlling the day to day life of the migrants have been reported in other studies (e.g., Horst, 2006). While many endure this kind of control, others break free by stopping communication altogether, as reported by Horst (2006), or imposing restrictions on family's communication with them, as explained by 26-year-old Bangladeshi worker Salim: "I [have] asked them [al]ready . . . *This time, this time you [can] call me, I know. This time this time my work time. Don't disturb [me].*"

In other cases, control being exercised via the mobile phone happens when the whole set of community relations in offline space is transferred to the interactions on mobile phone, as reported by Velayutham and Wise (2005). They observed that the Tamil-Indian diaspora from the villages observed the same kind of community relations while in Singapore, as they did in their places of origin. Moreover, the senior members who lived in the villages were eager to maintain the same hierarchical relations with the migrants. This resulted in village-based community members asking for money as a kind of reciprocal gift from the migrants. Selvan, who hailed from a small village in Tamil Nadu, India, told how he struggled to cope with the demands for money from family to neighbors and village acquaintances while being in Singapore:

> Anyone from my village would call up and request for monetary help. I cannot say "no." At home we provide employment for a number of people in our farms. They are given food whenever there is work and bonus on special occasions, apart from wages. So, all these people consider me as a rich and successful person. In the end, most of them are less eager to repay the debts.

Selvan's observations are typical of a social reproduction of caste relations on the mobile phone space. He is bounded by the norms of reciprocal gifting, and cannot escape the "higher caste" tag endowed on him. This controls his actions, responses, and approach to migratory life. Selvan might be enhancing his reputation among the village community due to his benevolence. But in the end, he is bounded by the nexus of relationship determined by caste norms and a general perception of migrants as always economically well off. Similarly, Landolt (2001) observed the bounded solidarity of migrants from El Salvador with the non-migrants, which provided enough room for enforceable trust. This translates to a kind of control enforced by the society on migrants: "In the receiving country and at home, their social lives continue to be so entwined with one another that those who do not send money to their families or do not *do right* by the community feel the consequences" (p. 232).

Malayali-Indian worker Jayan, 33, also had similar concerns. Because of this Jayan kept his mobile phone in his room at work. Jayan's apprehension was that any negative news from family over mobile phone, which includes a sudden demand for money in view of some expediency, would distract him at work and could lead to accidents.

As discussed under "management of everyday life," respondents had greater control over their own lives as a result of using the mobile phone. This is akin to the internalization of self-discipline observed by Foucault (1980) in modern societies. The technologies of the self are expanded to include mobile functionalities like the alarm clock, personal feature displays, account monitoring mechanisms, and navigation systems. These help discipline the migrant worker in the day to day workings of the modern industrial and knowledge economy. Migrants find their self-respect enhanced by contributing their part to the disciplinary mechanism, as control of oneself is rewarded and considered the best adjustment strategy in the host society. This control is all encompassing such that even pure enjoyment is rationalized as an optimal fusion of what the society offers and what one extracts for oneself. This is seen in the comments of Victor:

> There is life outside job. Certainly. If we work hard, we get good money. There is good [joyful] life for us in Singapore . . . it's not through work alone . . . Saturday and Sunday are off for me. That time I go with my friends for outing. To Universal Studio, St. John's island . . . we go these places . . . take photograph . . . after taking photograph, this is important . . . after taking photograph, before we re-enter our house [after the trip], our photos will be there on Orkut or FB. Buddies do it at the bat of an eyelid ["tak-nnu"] . . . they go home and do that at once. If the camera is with me, I upload the photos immediately after I return. To upload the photos and see how they are received . . . that's pure joy . . . (Victor, 25, Malayali-Indian, Welder)

In another case, Bangladeshi worker Mahim, 29, explained his rational use of the mobile phone by pointing at the extravagance of his friends on frivolous issues like dating and drinking. By contrast, although his spending on mobile use was above average,[6] Mahim boasted of his self-control. Rather than entertainment, his extra spending was a deliberate and clearly thought out strategy of talking at length, "up to an hour a day," with family over mobile phone.

In sum, three different meanings of control in connection to mobile phone use were identified in the appropriation patterns of the respondents: control by employers, control by families, and control of migrants' own lives. Collectively they added a pragmatic dimension to migrants' "usage and handling" of the mobile phone that had implications for their acculturation.

PRESTIGE AND SOCIAL IDENTITY

By examining the question "what the mobile device did to the user," the study attempted to understand the "prestige and social identity" involved in appropriation. The question brought to fore the symbolic meanings associated with mobile phone use, in contrast to the pragmatic use patterns discussed in the previous section. The material shows that there were two dimensions related to prestige and social identity, namely the notion of "warm experts" (Bakardjieva, 2006, p. 101) in matters related to technology use; and, gifting of mobile phones—used and new handsets—to loved ones in migrants' country of origin.

Gaining Prestige as "Warm Experts"

Looking first at Bakardjieva (2006), a behavioral pattern typical of a local "warm expert" is seen in the material. She characterizes "warm experts" as those who have both relatively deep knowledge of the technology's features, as well as knowledge of the concrete local conditions in which a technology is adopted. Warm experts not only possess the technical skills with which to operate a technology, but they are also willing to share experiences and interests with new learners. Tech-savvy migrants, who were "warm experts," shared their knowledge on the advanced functionalities of the mobile phone with fellow migrants. The material shows that as a result, these migrants were accorded recognition and expert status in matters related to mobile phone use. For example, a 43-year-old Tamil-Indian, Afsal, achieved warm expert status among peers by helping them download software that supported vernacular fonts:

> I download it [the software for vernacular fonts] for them [my friends]. When he sees Malayalam or Tamil in his phone, he is surprised and happy. He tells all his friends. Then many start approaching me with the same request. I help them. It's just for friendship. Many do not know how to open internet on mobile, but they have the equipment with them . . . I have taught people how to do it . . . in the last one year, I taught at least 50 people. Did you notice, just now I received a call . . . his mobile has a problem . . . he cannot read Tamil on it. He said he'll be coming here. . . . (Afsal)

These "experts" often had no formal schooling, but they were able to acquire technological skills, and they in-turn passed on the knowledge to their "laggard" peers. Such episodes helped widen the use horizon of less technically adept migrants, while boosting their self-conception and positively affecting their social standing.

Thus, the material shows that the migrants were able to use technical mastery of the mobile phone as an element in their status repertoire. They invested themselves in examining these potentials and developing this knowledge, sometimes to the detriment of their work, as we discussed in an earlier section.

Gifting

The second way that prestige and social identity were played out is seen in the practice of gifting mobile phones to loved ones back home. This was widely observed in the sample, especially among those with low digital literacy and technological skills. Both used and newly purchased phones were gifted to family members, as seen in the comments of Ganesh, a 30-year-old Malayali-Indian worker:

> I gave her [my wife] the phone after engagement . . . it was a good model, with camera, video and features like that. After marriage I presented her another mobile handset, and recently another one . . . totally three. (Ganesh)

The practice of gifting was, in some cases, a well-recognized cultural characteristic. As Velayutham and Wise (2005) observed, even in transnational spaces, some close-knit communities replicated the moral economy of the caste groups in which they belonged. Migrants from a south Indian village in Singapore, for example, were bound by a "gift-exchange relationship" (Velayutham & Wise, 2005, p. 38), wherein the community back home always expected that the migrants would "present gifts. . . assist with cash to help needy relatives, offer disaster relief, contribute fund to village projects and assist with finance to build new houses" (p. 30). Velayutham and Wise (2005) argued that the "thankful indebtedness" (p. 37) that the migrants felt toward their community was inculcated through strict endogamy and inward-looking principles followed by its members over generations. In a similar manner, through gifting mobile phones, the migrant workers in the present study strengthened their bonds with their home communities. Conforming to these norms were an element in ensuring the migrants' prestige and social identity in the local communities:

> One needs to entice relatives and neighbors with all sorts of things, when going home . . . electronic goods, duty free cigarettes, or things like that. Otherwise, they do not give value to people like us. I was very keen about gifting and carrying goods in the beginning. These days I don't care . . . But still I have this feeling. (Selvan, Tamil-Indian worker)

Selvan had created an opinion about gifting from his nearly two-decade long migrant life in Singapore. Initially, he was very enthusiastic about gifting various items to relatives and neighbors, which included mobile phones. Currently, he is apathetic to the sense of prestige that gifting brings to him. Thus, Selvan's words indicate that endowment of prestige and social identity is a dynamic process, wherein the recipient also has the agency to accept, reject, or negotiate it based on perceived repercussions.

Another motivation associated with gifting phones is that it legitimated the purchase of a new device for the migrant. Our material shows that respondents often gifted their used handsets when they moved on to newer, more advanced mobile devices. This purchase/gifting practice can be seen as a two-sided prestige exercise. As seen in the symbolic cycle in the circular model, a circular pattern of gifting and identity formation could be delineated in this phenomenon: while gifting mobile phones accorded the migrants prestige and social identity in their home community, ownership of an advanced mobile device accorded them prestige in Singapore. Malayali-Indian worker Francis, 30, changed his mobile handset "about eight to 10 times" in his seven years' of residence in Singapore. Whenever a better model mobile phone was launched in Singapore, he bought it for his own use and sent his old handset to relatives or friends back home. Similarly, Malayali-Indian worker Manoj, 23, gifted mobile phones to his friends back home even if they did not make a request. Cristo, 35, a Malayali-Indian worker, also recalled how he bought a smartphone for his mother, who showed interest in it after seeing that everyone around her, including her less-educated neighbors are listening to songs and videos on the smartphones:

> She [My mother] was not very keen about it earlier. In fact, earlier she had been using her feature phone very reluctantly, with lot of effort. But in the last one year, everyone in her friends' circle started using a smartphone—neighbors, her church-going friends, everyone. Being a widow, she also had to cope up with loneliness. So, once she asked, "you see, the servant in the other home uses a smartphone. I have seen her with earphones plugged in during her rest time. How come?" Then she slowly started to think about buying a smartphone for herself. That's the motivation . . . sort of feeling that one has to keep up with one's peers. (Cristo)

Cristo later bought a phone for his mother and gifted it to her on his next visit. Cristo recalled that his mother learned to use limited number of functionalities in the new phone with a lot of effort. However, she was happy that she could "take this step." Cristo's experience shows how the recipient enjoys higher self-esteem and a feeling of pride and prestige by conforming to the unwritten rules of the community. A subjective interpretation of the changing technological conditions in the neighborhood, coupled with Cristo's benevolence allowed his aged mother to take this step. A somewhat different twist on this practice can be seen in the comments of Ramesh, a 28-year-old Telugu-Indian worker:

> After coming to Singapore I've bought two phones only. The current one is Nokia, small. Thirty dollars. Small one. . . . [The] big one . . . [I have sent it to] India already. That was a costly one. I gifted it to my mother. (Ramesh)

Ramesh's experience reflected another trend among some migrants: The trend of keeping a low-profile mobile device for themselves and gifting an advanced mobile handset to their loved ones back home. Unlike those migrants who always kept the most advanced mobile handsets for their own use, respondents like Ramesh did not stake their local prestige and social identity on advanced mobile phones as fashion statements. Rather, in his case, it seems that he put more weight in his family's estimation of his status.

META COMMUNICATION

As posited by the circular model, the two cycles of (1) "usage and handling" and (2) "prestige and social identity" are linked by meta communication. Respondents not only used a wide variety of functionalities of the phone, as noted above, but they discussed these uses at length with their friends. This developed into a kind of self-perpetuating discussion among friends, which modified itself along with new brands of phone available in the market and newer functionalities.

There were reflections from migrants on the numerous functionalities of mobile phone. As we discuss in a section below, one type of user experimented with the technological features of the mobile phone, and then reflected on the illegality of such alterations, in addition to the possible breach of contract with the service provider. They also reflected on how their innovations (e.g., that helped a user read a vernacular font on the mobile phone) were understood by their peers. These reflections were crucial in the appropriation process, as they represented a recursive, iterative, and mutually reinforcing process involving both actual technology use and the technology's impact on users' identity and prestige.

In some cases, the mobile phone and its diverse uses were the topic of urban legends, gossip, and the grapevine. James recalled how he formed an opinion about the host society via its seeming fetishization of technology:

> When the iPhone was launched . . . girls camped in front of the retail shop throughout the night to book one piece for them. The next day's newspaper carried this photograph . . . iPhone is a craze here [in Singapore]. (James, 36, Malayali-Indian worker)

Similar to other research (Vacaru, Shepherd, & Sheridan, 2004), respondents also noted how use of the mobile phone could be a shield that people used to guard against social interaction. For example, James noted that people seemed to "hide" behind their phones on the buses perhaps to avoid engaging in conversation:

> Mobile phone diffusion is all good. But have you noticed the problem here? No one, no girl will look at your face when you travel in a bus . . . everyone is looking at the screens of their mobile phone. (James)

The material in our interviews shows how the interviewees interpreted these experiences in light of their self-definition and their position in the host society.

Reviewing the sections on "usage and handling," "prestige and social identity," and "meta communication," we explored mobile phone appropriation based on the circular model. The dimension of "usage and handling" explained the functional uses of the phone while "prestige and social identity" focused on what the mobile device meant for the users and how they used it in ways that affected their sense of identity and self-esteem. "Meta communication" linked the complex process of functional use of the technology with the effects of the technology on individual's sense of identity, prestige, and self-esteem. Based on these analyses, we now turn to the development of four different mobile phone appropriation types identified among the migrant workers.

MOBILE PHONE APPROPRIATION TYPES

We now turn to a taxonomy of users. The reader will recall from chapter 3 that we used the term appropriation in the sense developed by Bar, Pisani, and Weber (2007, p. 3) to describe how users go through a process of making a technology "their own." The concept is broader than simply purchasing the item. It also includes bringing the item into the flux of their everyday life. We used pre-determined concepts as well as categories that emerged in the analysis to identify these types. As noted in table 10 in appendix A, functional use, symbolic use, use of advanced features, cost considerations, and content of meta-communication were the categories associated with mobile phone appropriation.

We draw on Weber's (1904b) "ideal types," which consider these as *Gedankenbilder*, or "mental images." These mental constructs allow one to distinguish between important dimensions of a phenomenon. Following Weber, the notion of an ideal type is intended to provide broader insight into the characteristics of a population: "An ideal type is formed by the one-sided accentuation of one or more points of view and by the synthesis of a great many diffuse, discrete, more or less present and occasionally absent concrete individual phenomena, which are arranged according to those one-sidedly emphasized viewpoints into a unified analytical construct" (Weber, 1904/1949a, p. 90). Other characteristics of ideal types are adequacy, lack of correspondence with empirical reality in the sense of a positivist science, and priority for subjective ethical considerations over objectivity.

The types we introduce below have epistemological roots in Weberian ideal types. Although we present examples of individuals based on their characteristics to substantiate these types, the individuals themselves are not members of any type we introduce, but they merely embody the characteristics represented in their abstractness.

The Convenience Seeker

The first type of user to emerge from the data we call a *convenience seeker*. The convenience seeker compared the functionalities of the mobile phone with other media, and held an informed opinion on the complexities involved in adoption and use of a particular device. Convenience seekers had a more or less clear understating on which model to buy. This understanding was not intuitive, but an informed decision they made after careful consideration of factors including economic resources, their skill level, and their potential needs. Convenience seekers were not interested in the aesthetic dimensions of the mobile phone, or in other words, the symbolic effects of mobile phone were not of importance to them. Consequently, they were uninterested with praise or criticism by friends that went along with the possession of a particular mobile device. As satisfaction of basic utilitarian needs was paramount, additional qualities and functions were superfluous. The aesthetic qualities of mobile phones, for example, as a fashion accessory, did not appeal to the convenience seeker. What mattered to them the most was its functionality and utility.

A case to cite from the sample will be that of Lawrence. For him, the utility of the device provided the paramount focus. During the fifteen minutes' commuting time, he used songs on his phone to fill the time. This commuting time often coincided with the FM broadcast of vernacular songs. Thus, mobile phone was the best method to tune in to that station, while traveling. In the evening, after returning to the dormitory, he used a laptop computer to provide entertainment. But why this switch between different media devices? Can't there be an overlap of various features, affordances on these two devices? Lawrence replies:

> No, I don't use internet on mobile phone . . . There hasn't arisen such a need for me. . . . I know how to do it . . . we have to call them [the service provider] and open [activate] internet in our account . . . It's not at all difficult. But I don't feel there is a need. Firstly, it is a small screen . . . then I have the laptop now. So, since I have this facility already, there is no need.

This starkly contrasted with another type of users we discuss below who always needed expert advice for their problems related to mobile phone. In contrast, Lawrence sought ready optimal solutions. But for him it was not a question of showing anyone that he had achieved something, but only a matter of convincing himself that he had the most desirable solution to the problem at hand.

Theoretically, convenience seekers can be compared with *bricoleur* in cultural anthropology (Lévi-Strauss, 1966), to whom the task at hand is more important than the tools used for solving it. In the case of Lawrence, his objective was to find cost viable solutions to his requirements, that is, to be able to listen to music while traveling, to browse internet without hassle, and so on. He did not pay much attention to the devices that helped him achieve this end, but was focused on the result. Wirth and others (2008) also identify a similar type of mobile users in their study, which they call "Everyday-Life-Managers," who "emphasize(d) the pragmatic use dimensions of *control* and *management of everyday life*" (p. 608).

The Experimenter

The second type of user to emerge from data is the *experimenter*. Experimenters were not much bothered by the cost of either the device or the cost of use. Their knowledge about technical features were above par compared to others in the sample. In other words, experimenters had clear everyday knowledge of mobile phone functionalities. That said, they were not satisfied with this. Experimenters were positive skeptics. In order to find answers to the questions that arose in their minds, experimenters kept on exploring advanced and unique uses of the device. In the end, they came up with unique uses of the technology that went beyond normal use. For example, Viswanath, a 26-year-old Malayali electrician who worked in a shipyard, found out that his SIM card could provide connectivity to both his mobile phone and his laptop:

> I improvised on it one day. This is a SIM and what goes in here [in the laptop] is a SIM. So, there is no question of illegal use. So, if one's mobile phone supports this kind of SIM, then for them [the service provider], whether it's used in computer or in a mobile phone . . . they don't have to worry about it. For them they earn SGD 19 a month. If it's used in mobile phone, maximum 2GB usage will be there in a month. Maximum. But if we use in laptop, we can utilize more.

Viswanath was unsure whether the solution he had found would amount to breach of agreement with the service provider. Interestingly, he refers to this in terms of its legality, not simply as a breach in the conditions of his contract. Nevertheless, he rationalized his act using a technical logic. This ability to find a technological solution to an everyday problem characterizes experimenters.

In another case, Afsal, a Tamil worker, overcame his frustration over lack of English skills by downloading apps that could help read and input vernacular text in the mobile phone. He learned, to his disappointment, that the model he had just bought did not support vernacular font. As changing the phone was not a feasible solution at that point, Afsal explored a technological solution to the problem. And his efforts paid off:

> You know one thing, when you buy this model—5800—from shop, it can afford only a few systems. This 5800 model. . . in this mobile if you want to read Tamil or Malayalam, it is not possible . . . I mean when you buy from shop . . . But there is one software, if you download that software, then whatever you want to read—whether Malayalam or Tamil—you can read that. Without downloading the software, the font will not work. Sometimes, I do it for others . . . for whichever 5800 touch phones . . .

In the case of both Viswanath and Afsal, they had only a minimal reliance on external support. There is some justification in arguing that these workers' technological prowess with mobile phones was supported by the skills that they had acquired by dealing with other electrical and mechanical equipment on a daily basis in their jobs.

According to our material, the experimenters were seen as a resource on technical issues. They were often approached by their less proficient migrant friends for advice and help. Experimenters imparted their acquired skills and knowledge to others, which helped them gain a kind of "wizardry" status—somewhat like Bakardjieva's (2006) "warm experts"—in the community. Their advanced technical knowledge on mobile devices distinguished them from common users. However, experimenters still immersed themselves in the activities of the common users, while sharing experiences, interests, and knowledge with them (Bakardjieva, 2006, p. 101). This helped them mediate between the universal features of the technology and the everyday needs of common users situated in concrete local situation.

Experimentation with mobile phone affordances was found to have a spiraling effect. Initial success in learning led to greater investments and risk-taking. Experimenters perceived a kind of autonomy and self-empowerment in mastering the advanced mobile phone functionalities, which helped them psychologically overcome, or at least mitigate the effects of, their marginalized existence:

> When one becomes confident of searching knowledge on internet, there is no need to worry that "I have only pre-university education." We can learn anything on our own . . . move to places like Australia or even America. (Victor, 25, Malayali-Indian)

However, experimentation was not without downsides. Exploring the affordances of smartphones while at work led to problems. The experimenters sometimes tested this boundary, as noted by Malayali-Indian worker Deepan above. He noted that workers became inattentive towards the safety aspects of work when they were glued to their smartphones. Over-enthusiasm with the technology device led to over-indulgence. This eventually led the management to impose restrictions on mobile phone use, including penalties. As Bakardjieva (2006) observed, these respondents were "losing [themselves] in the technology much in the same way amateur musicians and drug addicts were submitting to their passion" (p. 104).

The Group Communicator

Group communicators were the social hubs of their respective groups. This group used the mobile phone to reach out and coordinate with their friends using the swiftest and most cost-effective methods. At the time of data collection, texting (that is SMS) was widely used within friendship circles and in, for example, religious groups. Group communicators identified themselves so much with the groups they were part of that it was quite impossible for them to consider mobile phone as a personal medium. When they recollected their mobile phone use, it was almost always in the larger context of communicating with their group. Due to this, they did not have clear notion of the amount spent regularly on mobile phone use. At least in some cases, the expenditure for mobile phone use was covered by group funds:

> I realized the importance of text messages fully after coming here [to Singapore]. How many calls can we make to the members? There are many members . . . For a message, it's easy, put all numbers in the distribution list and send. (James, Malayali-Indian)

James was part of a religious group that met regularly for prayer meetings and recreational events. It may be noted in passing that the group in which James was part of was limited to low-skilled workers. Professionals and newly naturalized citizens or residents of the same ethnic community—Malayali—had other groups to socialize, although they belonged to the same religion. This exclusivity of groups was observed in regular prayer meetings and other recreational events. So, James, on behalf of the religious group, was actually sending the text messages to fellow migrant workers in the same class. Being in a class with limited pooled resources and limited access to other forms of App-based communication platforms,[7] James had to economize on the communication expenses in the interest of everyone in the group. Thus, he chose to send text messages in bulk. In fact, it was not James's original idea to send bulk SMS. But the point to note is that he understood the importance of this practice and followed it without fail when his turn came.

Group communicators had membership in a range of overlapping groups; for example, members of a prayer group might be coworkers in the same work organization or inmates of the same dormitory. Group communicators often assumed leadership roles in work organizations or social congregations bound together by ethnicity, politics, or religion. Those who were interested in the politics of their home country kept detailed track of national political events. Our material shows that some were even engaged in vote canvasing via mobile phone, and actively utilized it to give strategic directions to their co-party workers (Aricat et al., 2015):

> At regional level elections, more than using mobile phone to receive information, I have used it to make people think favorably towards certain candidates. . . . e.g. one person [candidate in an election] in my village called me and requested me to help him. Since I knew about him from the past, even above politics, I tried to influence people at the time of elections using mobile phone . . . thinking that a particular candidate was good . . . Specifically, I called my friends [in India] who would listen to me and asked them to help him [in the elections]. (Kesavan, Malayali-Indian)

Theoretically, the characteristics of group communicators can be explained using the concepts of social capital, and opinion leadership. Social capital, according to one definition, is "a resource that actors derive from specific social structures and then use to pursue their interests; it is created by changes in the relations among actors" (Baker, 1990, p. 691). It is the positive aspect of social capital that finds emphasis in the case of group communicators, as they work towards enforcing a norm of reciprocity in the group. Opinion leadership is a key concept in diffusion studies which posits that change and innovation do not reach the common audience directly, but are mediated and interpreted by the opinion leaders or influential members in a group (Dubois & Gaffney 2014). Here, the group communicators not only act as an initial node from which a mass message emanates, but they also do a kind of gate-keeping and mediation of the messages, making them more meaningful and relevant to the group.

The Austere User

Finally, as the name implies, *austere users* describes those who are content with their low-end phones. This group often had little in the way of economic resources. In addition, they also lagged behind in motivation and in technical skills. This further limited their access to more advanced features of the mobile devices. This group of users learned basic mobile phone skills from others, often with some difficulty. Other than that, they had no interest in exploring the advanced functions of the device. Having only limited needs of communication, austere users did voice calling and texting on a feature phone. They had an occasional need to take a picture and play music. In contrast to experimenters' technology obsession, the austere users desired for fewer complications in technology use, and wanted to avoid unnecessary technological clutter:

> I would like to buy an advanced mobile phone to listen to songs in good quality sound; camera, with good clarity. But since my salary is less . . . My present mobile phone has camera, but it doesn't have that much clarity. Other features that I intend to use. . . . no nothing particular [dismisses any chance of learning to access internet on mobile]. (Kannan, 23, General Worker, Malayali-Indian)

> I've no such desire [of accessing internet on mobiles] . . . no thoughts of doing that. Because there is no need for that . . . they [my family] are calling me if required, and everything is going well. So no such desires. Whatever is happening, is because of God's grace. (Chandran, 45, General Worker)

Both Chandran and Kannan exemplify two important facets of austere user. Kannan's lack of enthusiasm to learn internet skills was driven by his lack of economic resources. Chandran, on the other hand, was generally satisfied with the scheme of events and thus lack motivation to learn anything more on communication technologies. In both these respondents, there was an underlying tendency towards status quo and hesitance towards accommodating anything that is in dissonance with their current world outlook.

Austere users had a tentative relationship to the technology. Use of basic features such as calling and texting on a simple feature phone was approached with much care and caution. These users would think twice before proceeding from one screen to the next. Indeed, they seldom strayed from the basic functions into a fresh learning experience.

Theoretically, austere users have parallels with "laggards" in the diffusion of innovation literature (Rogers, 2003). But their laggardness needs to be socially situated, given that they have comparatively low economic wherewithal, low social standing in the receiving country, and lack of human capital for social mobility.

SUMMARY

As a first step towards answering the research question "How do acculturation types relate to mobile phone appropriation types among migrants, when defined in non-essential and non-universal terms?" we first analyzed the data based on the circular model of mobile phone appropriation. The reader may recall that the circular model integrates "functional uses" of the phone with the "symbolic functions" that are ascribed to the ownership and use of the technology. Meta-communication, or communication on communication, provides the link between the two aspects of the appropriation process, which are "functional uses" (or, "usage and handling") and "symbolic functions" (or, "prestige and social identity"). Further, as a step towards achieving greater order and comprehension in data, we identified four ideal types of mobile phone appropriation. These were the *convenience seeker*, the *experimenter*, the *group communicator*, and the *austere user*. These types of users differed in their economic resources, motivation, and skills to move on to advanced features of the phone or Web 2.0, the recognition they have among peers and within informal groups, and their ability to improvise and innovate on issues related to technology. The next step is to identify acculturation types, which is followed by linking the appropriation/user types with the acculturation types. These are attempted in the following chapter.

NOTES

1. The situation was fast changing at the time of fieldwork, when respondents were increasingly moving towards data connections, smartphones, and internet-based phone applications like Skype on their mobile phones.

2. Incoming calls are generally charged by all three mobile service providers in Singapore. The amount varies depending on the plan, and the type of connection—prepaid or monthly billing. Further, for prepaid connections, all plans usually provide a bonus amount that can be used for both outgoing and incoming local calls. This local call bonus is usually two to four times the amount of the top-up. Thus, a Singtel top-up of SGD 15 has a validity of 50 days, and allows for SGD 15 for international calling (at subsidized rates for selected countries in Southeast and South Asia) and an allowance of SGD 40 for local calls. Any unused amount will be carried forward, within a maximum limit, if the user tops up again within the validity period.

3. Indian sub-ethnicities are named after the languages they speak. A Malayali belongs to the southwestern state of Kerala and speaks Malayalam, a Tamilian belongs to southeastern state of Tamil Nadu and speaks Tamil, and a Telugu belongs to the southeastern state of then undivided Andhra Pradesh and speaks Telugu. Bengali speaking communities straddle the northeastern Indian state of West Bengal and Bangladesh. Barring three, all other Bengali speaking respondents were from Bangladesh.

4. Being stereotyped is not something unusual to immigrant minority communities. On stereotypes attached to low-skilled workers, see Hamid (2015), Ryan (2010), among others.

5. Indeed, a non-call cannot be defined using the verb "to be." Here, "non-calls" refers to any trace of the respondent related to mobile communication that can "present" him to loved ones. It can be the respondent's contact number stored in a family member's phone, a missed call from him to the family, or even a trace in memory that anticipates a call from him at a particular time.

6. As a worker in the income bracket of SGD 500 and 999 a month, Mahim's monthly expenditure of SGD 120 was higher than average for a worker, as he himself admits.

7. In 2011–2012, the smartphone-based applications for messaging were not common among migrant workers. The SMS may have gone obsolete over the years, as instant messaging WhatApp became popular among this group of migrants. However, the characteristics of the "group communicator" and the motivations behind group messaging have remained more or less the same.

Chapter Seven

Acculturation and Adaptation

In the previous chapter, a detailed picture of migrants' appropriation of the mobile phone was drawn. We elaborated on the functional uses of the phone, the symbolic meanings users derived as a result of their interaction with the technology, and meta communication that linked these two dimensions of appropriation. The chapter also sketched out four types of users based on their mobile phone use.

This chapter extends the investigation into the second major area that is explored in this book, which is acculturation. This chapter examines how migrants' use of a mobile phone is related to their adaptation to work as well as to their social life in the host country. In order to have a quantitative understanding of the acculturation phenomenon, we followed the bi-dimensional acculturation model developed by Berry (1997). The survey responses provided insights into (1) acculturation, (2) mobile phone calling to dominant host and home cultures, and (3) migrants' adaptation outcomes in social life and work organization. Analysis of this material is followed by a qualitative exploration of the phenomenon. The chapter concludes by drawing links between acculturation and appropriation.

The first question that we addressed asked, "How do cultural identity, multiculturalism in workplace, and mobile calling to home and host cultures affect migrants' adaptation outcomes in the host country?"[1] Adaptation outcomes were divided into two domains (i.e., related to work and social life). Consequently two main hypotheses were formed, one each for each dependent variable. Finally, statistical tests were conducted to test these hypotheses.

There was a mixed pattern of relationships between the independent[2] and the dependent[3] variables. As the reader will see, age, education level, and multiculturalism in workplace had positive influence on migrants' adaptation to social life. However, no other independent variable had effect on adaptation to social life. In the work domain, three independent variables—cultural identity, multiculturalism in workplace, and mobile calling to dominant host cultures—emerged as significant predictors of the dependent variable organizational commitment. We have further used interview data to examine the various dimensions of these findings.

ADAPTATION TO SOCIAL LIFE AND ORGANIZATION: HYPOTHESES TESTING

As noted, the first hypothesis stated "multiculturalism in workplace, mobile phone calling to dominant host cultures and mobile phone calling to home culture positively relate to migrants' adaptation to social life, whereas cultural identity negatively relates to migrants' adaptation to social life." To test the hypothesis, a hierarchical regression was run.

Among demographic variables, age and education significantly predicted migrants' adaptation to social life.[4] The results indicate that those who are older and have higher educational qualification tend to be better adapted to the host culture (see table 8 in appendix A). Among the factors related to acculturation, multiculturalism in workplace significantly predicted migrants' adaptation to social life.[5] The results indicate that migrants who work in multicultural teams, and who get along well with colleagues from different cultures, are likely to adapt well to their life outside the organization. Moreover, migrants who are favorably disposed to other cultures in the workplace are more likely to adapt to social life in the host country. Among variables on mobile communication, none was a significant predictor of migrants' adaptation to social life. This means that no matter how long they talked to family or members of other cultures, it neither improved nor hampered their adaptation to social life.[6] There was also no interaction between acculturation and mobile phone use, implying that when a particular acculturation attitude affected their adaptation, amount of mobile phone calling could not alter the impact.[7] The hierarchical regression explained 8.9% of the total variance. This is rather low in social science research, and points to the existence of other factors relevant to social adaptation that are not included in this model.

The results indicated that, more than any other factor, multiculturalism in the workplace is the predictor of social adaptation of labor migrants (H1a is thereby supported). Neither mobile phone communication to host and home cultures, nor their strength of identity towards their own culture hampered their adaptation to social life (H1b, H1c, and H1d were thus not supported). Considering that migrants' adaptation to the social sphere of the host country is well above the average (see table 2 in appendix A), this result modifies current understanding of the role of mobile communication among labor migrants. Irrespective of the extent of migrants' mobile communication, their adaptation to the host society remains high. The results suggest the need to review the hype surrounding the impact of mobile communication on migrant adaptation.

The second hypothesis asserted that "multiculturalism in workplace and mobile phone calling to dominant host cultures positively affect migrants' organizational commitment, whereas cultural identity and mobile phone calling to home culture negatively relate to organizational commitment." To examine this we again used a hierarchical regression analysis.

Going carefully through the findings, the analysis shows that none of the demographic variables accounted for any significant amount of variability in the organizational commitment.[8] Thus, age, education, income, or period of residence did not affect migrants' level of commitment to organization. Acculturation was a significant predictor of organizational commitment, as both multiculturalism in workplace and cultural identity emerged as significant predictors.[9] The results indicate that migrants who are favorably disposed to other cultures in workplace, and who show greater identification with ethnic culture are more likely to show greater commitment to their organization in the host country. Mobile phone calling to members of the dominant culture was a significant predictor of organizational commitment.[10] This is substantiated by interview data that we discuss below, which, in short, shows that migrant workers increasingly use mobile phones to coordinate work with team mates, resulting in greater involvement in work and increased commitment to the organization. The interaction between multiculturalism in workplace and calling to members of the host culture was significant.[11] This implies that in the case of those migrant workers who are open to multiculturalism in organization, mobile communication to dominant host cultures increases their organizational commitment. The hierarchical regression model explained 43% of the variance in total, which is relatively strong considering the standards of statistical models in social science research (see beta coefficients on model in figure 4 in appendix B).

In view of these findings, hypotheses 2a and 2b were supported. This suggests that the positive relations between (i) multiculturalism in workplace, and (ii) mobile phone calling to host cultures, and organizational commitment is supported by data. Hypotheses 2c and 2d were not supported, suggesting that (i) mobile phone calling to home culture, and (ii) cultural identity, did not have the effects as hypothesized on organizational commitment. It is interesting to note that cultural identity was expected to negatively relate to organizational commitment. However, the results indicated significant positive relation between cultural identity and organizational commitment. This provides some insight into the nature of labor migrant integration in Singapore. Although the labor migrants identify with their cultural heritage, this has not hampered their adaptation to work. In fact, their cultural identification has only enhanced their adaptation to organization.

The following sections triangulate the quantitative results with interview data, providing further support to the relationships identified between different variables in the statistical model.

ADAPTATION TO SOCIAL LIFE AND ORGANIZATION: EVIDENCE FROM INTERVIEWS

Interviews with the respondents[12] further supported the quantitative results noted above. Three factors—cultural identity, multiculturalism in workplace, and mobile phone calls to members of the dominant host culture—were positively related to organizational commitment. The relationships were irregular and the patterns were mixed in the domains of work and social life. However, to make a broad observation on the results—migrants' work and social life overlapped with each other, such that adjustment and contentment in one domain helped improve adaptation to the other domain. On the other hand, if the going on was bad in one domain, it threatened adjustment to the other. The following sections further explicate these points.

Commitment of the Workers to Work and to the Company

The reader may recall that we adopted a psychology-based definition for organizational commitment. Three constructs—compliance, identification, and internalization—characterized the concept of organizational commitment. Literature from business communication has largely supported the idea that better communication between employees helps improve employees' satisfaction and commitment towards work. Our material showed that a large part of the communication between coworkers at work was channeled through mobile phones. This facilitated their work, and helped develop a positive attitude towards work and organization.

Deepan, 40, a Malayali-Indian, worked as an air-conditioning mechanic in the shipyard. His daily task was to take measurements of air-conditioning pipes and fittings, and to arrange materials that were needed to fix any problem in the system. While carrying out the task, Deepan had to use his private mobile phone to talk to his supervisor and site engineer:

> I use mobile phone during work to discuss work related matters with coworkers and company professionals like supervisor and engineer. Calls are related to

work, like arranging materials . . . in my case, my duty is to take measurement. So, on the lines at which we had work, I take measurement and have to arrange materials for those lines. For matters related to these, we call the supervisor and engineers always on mobile phone. Generally these people will be away. . . . They'll be in the work site, which is less than one kilometer. We use mobile phone to contact them. Sometimes, when they're in office, their superior officers may be calling them and enquiring about the finished work and instructing them to start the next work. Then they'll call and ask us to start work on a particular line. Without mobile phone this coordination will not happen. Otherwise they will have to come down to us and tell us about the work, face to face. That's quite cumbersome. They also expect us to have a mobile phone. (Deepan)

Deepan's comments illustrate how the device facilitates his work, and there is an implicit expectation that he have one (Stephens et al., 2017). As we saw in the earlier chapter, Deepan was among many senior workers who pulled double duty on their mobile phones, without being paid for the work-related use. In fact, most of the respondents had subscribed to a service plan that fully or partially subsidized the in-country and incoming calls. This helped them use their private mobile phones for work-related calls, while not being reimbursed for the latter.

In a study among brick makers in Myanmar by the two authors of this volume, a similar mode of communication between brick unit owners and truck drivers was observed. In one case, a truck driver used his private phone and paid for the calls he made to the employer, mostly in relation to logistics and coordination. His calls to the employer was brief, whereas he had the luxury of more airtime if the employer called him. When asked why he did not expect to be reimbursed for the work-related calls, the driver dismissed it as a "meagre" amount. It later emerged that this driver had a long-standing loyalty with the brick unit owner, as he had been working in the unit for about 14 years. Thus, the employee's bonding with the employer was considered to be beyond usual employer/employee relations moving over into a more social level of interaction. For example, the employer supports this driver financially during rainy season, when most of the brick kilns are closed and thus, he does not have an income.

Similarly, migrant workers like Deepan did not expect to be reimbursed for their brief in-country calls related to work. This is suggestive of their loyalty to the organization, and commitment to work. In fact, when workers talked about cultural adaptation, their experiences in social life and in organization were intricately blended together:

The culture here. . . . I have traveled and worked in four five Gulf countries. Compared to them, the culture here is good. Other countries are not like this . . . there are problems, many problems in other countries. . . . Then those with us, they

> also cooperate with us well. In fact everyone in this place is cooperating with me well, be it a Singaporean, Malaysian, or Chinese, all are cooperating [helpful]. . . . The work culture here is different [in comparison to that of other countries where I worked]. Although I did the same job that I learned in other countries, here [in Singapore] there are a lot of additional equipment. As a result, the work is also demanding and we feel tense at times. . . . In order to adapt to it, one takes some time . . . usually some time is given for making this adaptation. (Deepan)

Deepan's work history attested to the way in which his work facilitated his adaptation to social life. Comparing his life in Singapore with that of his earlier work stints, Deepan recalled that he worked in five countries in the Middle East in the past. But he quit the job in each country within six months after joining. On the other hand, he had been working in Singapore for four years. In addition to the "economic benefits" of his current job, he liked the work culture in Singapore that was, to some degree, facilitated by being in social contact with his workmates. This has resulted in his seeking an extended stay in the country, despite the limited opportunities available for naturalization to the country.

There was a kind of blurring of boundaries between work and social life: coworkers came together to complete a task during work, and this bonding was later carried over to their social life. Since most of the coworkers were friends even outside work, there was a tinge of informality in their communication even at work spaces. Adaptation to social life, thus, followed a complex process in which work-related mobile phone calls had a major role to play. Mobile telephony turned out to be an effective tool in their day-to-day negotiations across the domains of work and social life:

> I do communicate with [people from] other communities, especially at work—to Burmese, Bangla and Thai. I call them sometimes, outside work also. With some I do talk outside work . . . . Say, when he goes home for leave then also I maintain contact. (Palanivel, 26, Tamil-Indian)

To solve specific work-related problems, respondents communicated with supervisors, colleagues, and friends. This communication, although functional and goal-oriented in nature, also provided an opportunity for cultural learning. The benefits of such communication were extended to social life where it helped facilitate respondents' adaptation. However, to benefit from their multicultural engagement in work spaces, the workers needed to have a multicultural disposition. This is evident in Palanivel's recollection noted above. Many respondents still had the contact numbers of their earlier colleagues in their mobile phone. Even when they left the job and returned to their home country or eventually moved to a third country, they stayed in touch over social networking sites such as Facebook.

The material showed that there was a synergy between work and socially-oriented mobile phone use. When workers from different cultures came together in a team at worksites, they used mobile phones to micro-coordinate. This allowed them to complete the assigned tasks faster and with greater ease. Later, the friendship between team members was extended to the social domain.

The link between cultural identity, ICT use and commitment to work is seen when workers who shared ethnic identity were able to informally coordinate the completion of tasks via the mobile phones:

> I often call on the number of the storekeeper, with whom I have become friends, and say "such and such material is in short supply.". . . I just need to wait at the site, the material will come there . . . My supervisor supports this kind of coordination. . . I think the organization appreciates it. (Narendran, 28, Malayali-Indian)

> It is usually in construction sites that mobile phones help workers to coordinate. In other sites where there is prohibition on mobile phones, foremen carry *Grid*[13] phone, with or without SIM card, or walkie-talkie. When general workers coordinate work using mobile phone it actually helps. In some cases, the sub-contracting companies provide mobile phone allowance to selected team leaders in a construction site. But not to general workers, though. (Kesavan, 35, Malayali-Indian)

The comments of Narendran showed how he was able to leverage on shared language and ethnic background to micro-coordinate and complete professional tasks using non-formal channels. Previously, Narendran, who was a general worker in the Process Construction and Maintenance sector, had to make several trips from his worksite to the store to collect materials. Eventually, he befriended the storekeeper who was Tamil. Although Narendran's mother tongue was not Tamil, he knew the language due to his native town's geographic proximity with the Tamil-speaking State in India. Capitalizing on this acquaintance allowed him to save time on trips, as the order for materials could now be made over mobile phone.

Narendran's strategy resembles the economic uses of mobile phone that have received wide attention in *ICT for Development* literature, wherein mobile phones have helped businesspeople to cut the number of trips to procure and sell the products (Ling et al., 2017; Saunders, Warford, & Wellenius, 1994). The implication of this strategy in the context of migrant workers is that the workers' identification with their ethnicity and their co-ethnic collaborations have helped them perform better at work. Expectedly, when work was facilitated, their supervisors and senior members like Kesavan appreciated it. Kesavan, however, observed that even in construction sites where mobile phones were effectively used for work coordination, the sub-contracting companies reimbursed only the mobile expenses of the foreman or the team leader. Thus, Narendran's mobile-based coordination of work is an informal arrangement between workers, which nevertheless facilitated work. The supervisors approved of it, but the companies were indifferent to such practice.

Migrants' commitment to work was also intricately linked with their national identity. As soon as the migrants cross the border to a foreign country, the sub-identities are mostly superseded by a more overarching national identity. Indeed, a national identity more often than not helps them ignore, if not overcome, the discriminations they faced in home country as a result of their marginalized sub-identities. For example, those who were identified as lower-caste[14] in their countries of origin faced a reduced level of stigma in the host country.[15] The migrants then diffused the newly acquired ideas back to their home countries, allowing their countrymen to appreciate how the companies, the government, and the civil society in the host country perform with optimal results. Faizal, a social science graduate from Bangladesh, could only find a job that was far below his educational qualification. However, he treated his tough working conditions in Singapore as an opportunity to learn. In the interview material, he compared the living conditions in his own country with those in Singapore:

> I want to make them [my fellow countrymen] concern what's going on in my country . . . I make them aware [using the channels of internet and mobile phone]. I talk to them about these matters . . . Singapore rules and regulations . . . it's strict. But if somebody follows it . . . he can learn something from Singapore and use it going back to the country. I'm happy that I could work for this organization. (Faizal, 31)

Faizal's comments suggested that a commitment towards organization involved development of a new identity to the migrant workers. This identity provided them with a kind of authority to evaluate their experience in a comparative perspective and to provide opinion leadership to fellow countrymen back in their place of origin.

It is possible to read out of Faizal's comments a respect towards the host country's system of governance. This played through to his commitment towards the organization where he worked. Finally, bringing this around to the findings from the quantitative material, mobile phones, supplemented by internet, helped him disseminate this learning both from the host society and from the organization to his fellow countrymen. Faizal's national identity was also strengthened by feelings of social responsibility and civic sense. This led to his greater commitment towards his work organization in Singapore, as it gave him greater financial security. His sustained earnings in Singapore helped him fulfill his social responsibility by donating to the neighborhood associations and supporting the needy back in Bangladesh:

> Sometimes I have to excavate very dirty soil. Yet, I don't mind, because I have work at least . . . my country can never provide me the job that Singapore provides. [While being here] I am able to do something for my home country. Last time, with my friends, we took some decisions that some of the peoples in my country, they have no shelter. They have to lie down on the footpath. But if rain comes, they have to run away, here and there. I feel very bad about it. And that's why, voluntarily, I and all my friends took decision that . . . [we] will collect donations from workers of all countries here . . . some help. We'll directly send [the money] to Bangladesh. Mobile phones help me in all this, to co-ordinate with my friends here. To co-ordinate with people back in Bangladesh, the villagers back in Bangladesh. I am still in touch with people working in [voluntary] organizations in Bangladesh. I get to know about their daily affairs over phone. (Faizal)

In sum, workers who have a multicultural orientation find it easier to garner support from their teammates. We can see this in the comments of Narendran who calls the storekeeper and uses his acquired understanding of Tamil to resolve logistical issues. It is seen in the case of Deepan, who liked his job beyond the "economic benefits" it provided, mainly because of the work culture that Singapore provided. This is also seen in the experiences of Faizal who collected donations from workers from other cultures and showed commitment towards the organization in an attempt to negotiate and redefine his identity as an opinion leader. Mobile phones help migrants to coordinate work in teams, which most often extend to social life outside work. Following certain strategic uses of mobile phones at work, the migrant workers could complete the assigned tasks more quickly than the stipulated time. They gain the appreciation of the supervisors as a result, although they are rarely reimbursed for their work-related calls. The treatment of the company management, however, has not always been uniform for all workers. The following section discusses this.

Treatment of the Workers by the Company in the Era of Mobiles

Company policies regarding worker access to information vary. In some cases, there is only limited access to the communication for personal reasons. Personal information was also filtered through official channels in companies before the advent of mobile phones. In other cases, the companies were more open to this type of communication. Our material shows that these policies play through to the migrant's evaluation of the working situation.

Those migrants who have worked in other countries before moving to Singapore could recall their life experiences based on the different modes of communication that they used to communicate with families:

> When I was in Saudi [Arabia], I was not even informed when my father passed away. The company didn't inform me. Some firms are like that, especially in Arab countries. They don't inform, thinking that we will leave immediately. So, I got to know [about my father's demise] after a few months. Everything was through letters then. Now, we call daily. (Daniel, 38, Tamil-Indian worker)

While Daniel is content about the progress in communication technology, he is embittered by the company policies allowing the unjust treatment of migrant workers. The policies on migrant rights, and access to personal communication, vary by country. Following the experience of Daniel, these are the least supportive in the Gulf. In another case, a migrant worker recalled how the supervisor in Singapore cared about him when the news about the sudden death of his grandfather was passed to him:

> Once a call came from home saying my grandfather passed away. They [Kannan's family] did not call me, but my manager. The manager called to our work site and conveyed the news. He asked my supervisor to allow me to sit in the canteen for 10 to 15 minutes. He permitted me to call home, or to go back to my room, whatever I chose to. When I think about it now, I feel good about these allowances. The company has treated us well. (Kannan, 23, Malayali-Indian)

These examples provide an overall picture of companies' treatment of workers. They help understand the power relations inherent in the communication exchanges between migrant workers and employers.

Our material shows that mobile phones helped bring timely information and support to migrant workers in organizational and social spaces alike. Our interviewees expressed confidence that if they got into trouble somewhere in Singapore, they could call up their boss for support. Respondents who worked on and commuted to islands that were located away from the Singaporean mainland told of receiving the assistance of their superiors via the mobile phone. They said that they were sometimes stopped by the security guards at the gates. To gain entry they had to call their boss who alone could certify their identity before they were provided access to the worksite. Cases such as this show how the mobile phone facilitated the functioning of the organization and made the lives of the workers easier.

This type of support from organization sometimes took a personal turn, helping migrants to negotiate the challenges of cultural and social adaptation. Malayali-Indian worker Jayan, 33, explained how confident he was when his boss could be reached in case there was a problem even outside work:

> If we get entangled ["chikkiyal"] in to any problem here in Singapore, we immediately call up our boss and inform him. Nothing of that sort has happened as of now. But we have confidence that if we get into any trouble anywhere, we can call our boss. His number is with us. (Jayan)

While there are examples of mobile-based communication facilitating the situation of workers, there were also instances when hierarchical organizational structures and employee-unfriendly policies obstructed the communication between employees and the management. In such instances mobile communication acted as conduit of top-down communication from the management to the lower-rung employees. Respondents' commitment to the organization ultimately suffered in such cases. This can be seen in the example related by 23-year-old Malayali-Indian worker Vineeth. He reported that on occasions, he and his colleagues had questioned management regarding undue deductions in pay, less hospitable living conditions, and inconvenient transport facilities. In normal cases, dialogue with the management would help mitigate tensions. Vineeth reported that he felt that he could call his boss, who he referred to as "Madam," if there was a specific need. He had the sense that this line of communication could be used to resolve issues:

> We don't talk usually to anyone in the management, except to the madam. We call her when any need arises. Madam is in Malaysia . . . It's not as if we call daily, but only if we have a difficulty. (Vineeth)

The reliance on this communication channel, however, faltered on occasion. Vineeth noted the situation where one day, on their way to work in the shipyard, he and his colleagues were denied entry at the work site. Normally, the team leaders inform the personnel at the security gate as to the names of the work detail on a particular day. When Vineeth realized that this information had not been passed on to the security personnel, he called up "Madam" on his mobile phone in hopes that she could address the situation. Unfortunately, it was not possible to sort it out and the workers had to return to their dormitories. To compound the situation, management refused to pay that day's salary to the workers who had returned to the dormitories. Vineeth noted:

> She said, "[. . .] it will be marked a leave today." . . . We didn't get that day's pay. She is good for nothing. This company is no good. (Vineeth)

His comments indicate that in spite of a relatively open communication channel that can be used to resolve situations, management is not always responsive to particular situations. It may well be that "Madam" was not able on such short notice to address the issue. However, the episode shows that, at the point of departure, Vineeth conceived of the mobile phone as tool with which the situation could be addressed.

In summary, the interview material suggested that migrants formed an opinion about the ways in which companies treated them, which directly related to the expectations they had about the management. With the help of mobile phones, respondents gained access to managerial processes that had been out of bounds for them until then. With the help of mobile phone respondents could register personal grievances with the management. While many were contented by these possibilities, there were situations when the expectations of the migrant workers were not aligned with the actual outcomes. This mismatch resulted in a sense of disgruntlement among a group of workers. Mobile communication interacted in such cases to produce mixed outcomes of commitment towards organization and social adaptation.

Mobile Phones Failing to Break Cultural Barriers

There were situations where mobile phone communication with colleagues did not enhance respondents' interaction with other cultural groups. Interviews revealed the existence of instrumental, matter-of-fact engagement with other cultures, which did not go beyond the resolution of professional situations. Entrenched moral values related to the home culture put constraints on the migrant workers when they tried to gain membership in friendship networks. Malayali-Indian Victor, 25, was one among the early adopters of social networking sites like Facebook. However, he restricted his friends' circle on Facebook to members from his own ethnic community. There were indeed a few Singaporean "friends" in his Facebook list, largely because they were somewhat removed from him in their work-based organizational hierarchy:

> There is now a Filipino girl in our company. She's on FB [Facebook], but I've not added her as a friend. If one enters into that area, there is a chance to go in a wrong way . . . like having a girlfriend, and things like that . . . Since I don't like it, I'm not going that way. (Victor)

This shows how his anxiety with regard to eventual romantic entanglements meant that he simply avoided this type of social tie. It may also be that he feared a reprimand from members of his home culture or from family if he befriended a Filipino woman in Singapore. Similar to the experience of a group of south Indian migrants controlled by what Velayutham and Wise (2005) called a "transnational village gaze"[16] (p. 35), Victor's FB activities might be under constant monitoring by his family and friends. This supports the observation that migrants' identities are always a complex mixture of gender, class, and culture (Nazroo & Karlsen, 2003; Ryan, 2010). As theorized under "multiculturalism in workplace," the key task for an acculturating worker was to negotiate these identities in the organizational space and cope with the diversity among colleagues. The excerpts from the interviews suggested that the multicultural disposition of migrant workers could have a positive effect on their social adaptation. The comments also suggest that the mobile phones facilitated the process. However, there were aberrations to this common pattern, especially when migrant workers conformed strongly to the home culture and its norms, as evident in the example of Victor.

To summarize, this section explored relationships between migrants' (i) multicultural disposition in workplace; (ii) mobile communication with colleagues; and (iii) cultural identity, with outcomes of (i) organizational commitment, and (ii) social adaptation. The interview data revealed the mixed and complex nature of interactions between these variables. A broad conclusion that can be drawn from these results is that communication over mobile phone helps overcome the boundaries of work and social life. A migrant worker benefits if he easily adapts to work, and can go in the other direction if he faces discrimination or other kinds of injustices at work. The following section expands this investigation to the cultural sphere.

LINKS WITH HOME AND HOST CULTURES

We now turn to examining the data on acculturation, mainly under the rubric of how the migrants strove to retain identity while also relating to other cultures (see table 10 in appendix A for the themes identified under acculturation). The dimensions "retention of ties with the home culture" and "relating to dominant host cultures" are the cornerstone of acculturation. Migrants continually have to negotiate these two dimensions in their daily lives. In this analysis, we follow the same principles of taxonomy that we followed in the case of appropriation types in chapter 6. We explained in chapter 4 how Berry's (1997) four-type model, categorized acculturation in rigid categories of *assimilation, integration, marginalization,* and *separation*. The acculturation types we identify in this section overcome the limitations of Berry's model as they are contextualized for the population under study, that is, migrant workers in the low-income category. This section concludes by identifying relationships between mobile phone appropriation type and acculturation type.

Enduring Ties with the Home Culture

The material shows that retention of ties with the home culture was achieved in a variety of ways. Migrants' communication with family and friends "back home" helped retain the ties. Their participation in religious and social activities along with co-ethnics in the host country also helped strengthen their ties with the home culture.

First and foremost, the mobile phone was used to maintain contact with loved ones in the country of origin, as several examples in the previous sections have suggested. Migrants' interaction with co-ethnics in the host country varied according to ethnicity and religion. For instance, our informants included members of a small Malayali Christian prayer group that met regularly. Under the aegis of this group, the members reached out to the needy, cutting across ethnicities, to organize recreational and charitable activities. James, a 36-year-old from Kerala, noted, for example, that this group, of which he was a member, had organized a visit to Little India in 2011, with the objective of reaching out to the needy and the destitute. That event was seen as being successful, but they found that it was difficult to sustain the original enthusiasm:

> "Teka[17] mission" was a famous project we launched . . . we reached out to Indian workers in Little India. . . . but later people dropped out, mainly because of work pressures. (James)

The transient nature of such networks meant that the members failed to sustain momentum. The enthusiasm among the members ebbed and flowed.

In other cases, migrant groups were demarcated based on the professions and class characteristics within the Tamil or Telugu community. This, in turn served as a way to retain the vitality of their cultural as well as their "worker" vis-a-vis "foreign talent" identity.[18] There were, for example, separate groups for low-skilled migrant workers, nurses, IT professionals, and so on. Thus, the interactions between co-ethnic migrants were focused on the class-based group in which they were members. Unlike the more ad hoc activities described by James in the previous paragraph, those respondents who were associated with well-established social welfare organizations, especially those supported by resident Singaporeans, enjoyed more sustained social activities. This can be seen in the case of the Malayali-Indian worker Pavithran, 42, who volunteered at an old-age home in Singapore that housed Indian residents:

> When in India, I had been active in this mission [for the elderly]. I came here, I inquired about it. . . . I go there [to the old-aged home] on weekly off days. (Pavithran)

Unlike the informal and provisional activities of James, the stability of the old-age home meant that Pavithran could make a long-term engagement with its activities. His comments also indicated that he shared his participation with others, perhaps to recruit others, but also to mark his commitment.

Another way to maintain ties with the home culture was to access vernacular language news. The respondents reported that they could get information from home via mobile-mediated conversations with family and friends where they could attach links to relevant digital items and also stream video. Further, the interviewees noted that they could contact friends and family back home to gain better insight into the relevant news items.

This is seen, for example, when elections to Tamil Nadu State legislative assembly were held in India. Hussain, 25, a Tamil-Indian worker, followed the vote counting process via television in Singapore. When he started to doubt the fairness of election process in one of his home state electoral constituencies, he called up a political contact in the constituency who confirmed that there was widespread vote rigging in the constituency. This low-threshold to telephonic contact meant that Hussain was able to get first-hand reports. In addition, the secondary consequence of the telephone conversation was to refresh the link between Hussain and his local friend. Daniel succinctly expressed similar sentiments on accessing news from his locality:

> We get to know happenings not only from home, but even in the neighborhood . . . everything . . . someone falling sick . . . This is beyond what we get from newspapers. We don't get such local news from newspapers, right? (Daniel)

Thus, in addition to connections with family and friends, informal co-ethnic networks, voluntary organizations, friendship circles, and religious gatherings were important avenues that helped migrants retain a relationship with home culture. In some cases, the transient and informal nature of such networks limited their efficacy. Nevertheless, the material shows that the informants were active in seeing out through which to maintain long-term engagement with members of their home culture.

Relating to the Host Culture

In addition to being in contact with their co-ethnics, the material shows that the workers also sought out points of contact with members of the host culture. That said, respondents' engagement with dominant host communities varied across ethnic groups. Tamil respondents, for example, agreed that their acculturation to the host society was facilitated by the official recognition given to Tamil ethnic community in Singapore.[19] Tamil-Indian worker Ilavarasan noted:

> Mingling with the culture is not that difficult . . . It's easy. Tamil language is there. As you would have noticed, Tamil culture is there among the resident population, so it's easy for our mingling. (Ilavarasan, 28)

The historical ties between Tamil groups and Singapore played through to a lower threshold for interaction. Our material shows, however, that there were major differences between the settled Tamil-population in Singapore and the new Tamil immigrants from India. Similarities included shared religious beliefs, festivities, and language. However, Ilangovan, a 34-year-old Tamil-Indian, noted that the Tamil television channels based in Singapore cater to the tastes of the older population, airing films from the 1950s. Others also observed the differences between old and new Tamil immigrants in ways of dressing. Most of the peripheral cultural markers helped new immigrants to adapt with the host country as a newcomer. However, class barriers between the old and new, mostly low-skilled immigrants, made it difficult for the latter to find their place and to experience social mobility.

A barrier to contact with the dominant host cultures, as noted by some informants, was that Singaporeans classified low-skilled migrants as falling into a broadly defined class of "low income population," regardless of their ethnicity. As noted by Palanivel, a Tamil-Indian worker:

> Singapore culture is OK. Each one will have his own opinion about it. What I feel is that those in high status do not have any connection with us. They are in their own level. People like us are in our own groups. (Palanivel, 26)

This is a recurring theme in the literature on immigration in Singapore, especially associated with the situation of low-skilled immigrants (Hamid, 2015; Hua, 2014). Stereotypes like attribution of alcoholism, lack of civic sense, and unruly behavior among low-skilled male migrants are widespread. Adding to that, there is a pervasive language barrier among low-skilled migrants who have proficiency neither in Chinese nor in English, the two major languages of the resident population. Nevertheless, governmental agencies and NGOs have taken active initiatives to ward off residents' fear of the foreigners. Recreational activities involving both Singaporeans and the low-skilled migrants are carried out at work as well as in the communities.

Singapore's policy of promoting multiculturalism ensured hassle-free staging of cultural festivals, facilitating widespread participation of residents as well as non-residents in cultural festivals. Thus, migrant workers' engagement with the resident population provided a wide spectrum of experiences cutting across their social and professional lives:

> I volunteered for social activities . . . programs related to Onam [a regional festival of the community] . . . to the Malayali Association. We have gone to Yishun, earlier. Then to Timah [two places of Singapore] . . . all part of Malayali cultural festivals. (Bhaskaran, 50, Malayali-Indian)

Based on this complex picture on migrants' engagement with home and dominant host cultures, a typology of acculturation may be delineated. The types reveal how the migrants' engagement with home and dominant host cultures interact with their communication links with the two cultures and the multicultural context of the host society. This, in turn, is one element that plays into their acculturation experience.

Migrant Acculturation Types

Similar to the ideal types of mobile phone appropriation, the material suggests various ideal types, in a Weberian sense, when considering migrant acculturation. As noted in chapter 6, this includes what we have termed, the *culture campaigner*, the *culture connoisseur*, and the *cultural conservative*. The reader may recall, we introduced the bi-dimensional model as a first step with which to understand acculturation. This involved an investigation along two lines: (i) To what extent do immigrants want to establish relationships with cultures other than their own? and (ii) To what extent do they want to retain their own cultural heritage and identity? In chapter 4, we also introduced several criticisms raised against the bi-dimensional model for "essentializing" the concept of acculturation. We then noted an alternative method followed by diaspora studies, based on socio-historical understanding of acculturation, helping overcome the drawbacks of the bi-dimensional model.

The qualitative data opened up the dimensions of acculturation that are not captured by the bi-dimensional model. We used pre-determined as well as emergent categories to identify these types. As evident from table 10 in appendix A, five themes on acculturation emerged from the qualitative interviews. In addition, we followed the narrative presented by the respondents to identify intricate relationships between multiple categories, especially in the socio-historical context of Singapore. Although we present examples of individuals based on their characteristics to substantiate these types, the individuals themselves are not necessarily only members of any one type.

The Culture Campaigner

The first ideal type migrant we describe as *culture campaigners* were opinion leaders among their peers on matters related to their home culture. They identified with the home culture more than others in the community. In addition, they regularly promoted the ethos of their own culture, by helping organize cultural events that strengthen the cultural identity of the community. They took initiative to share culture-specific news with co-ethnics, and invited them to take part in cultural events. The example of Bhaskaran, a 50-year-old Malayali-Indian, above fits this category well. Bhaskaran received and circulated messages related to Malayali cultural events using his phone. He was an active advocate of the Malayali culture in Singapore.

Culture campaigners do not have any particular affinity or aversion towards the host country cultures. Indeed, they are mostly indifferent in this, perhaps because they have only a temporary status in the country. The example of Palanivel, a 26-year-old Tamil-Indian, who was described above, illuminates this point. Palanivel knows well that he belongs to a particular class of immigrants. Thus, for him the options/motivation to engage with the dominant host cultures of Singapore are limited. He fills this gap by actively engaging with his own Tamil community and campaigning for its cause. Most *culture campaigners* were active in ethnic, religious, or cultural organizations where their unique bonds could be maintained and developed in interaction with co-ethnics.

This group of individuals often had a higher level of literacy and was skilled in the use of advanced mobile handsets. This can be seen in the case of James, the 36-year-old Malayali-Indian, who we met in the last chapter. He had learned about the utility of mass text messages and used the feature effectively to notify the group members on weekly prayer meetings. Culture campaigners, thus, actively utilized the possibilities of technology in their advocacy.

The Culture Connoisseur

The second ideal type migrant is the *culture connoisseur.* These individuals were characterized by not wishing to cultivate that which was specific to any particular culture. Instead, they focused on orienting themselves across groups. Connoisseurs took the perspective of trying to understand the broader social trends and characteristics. Hence, when relating to Singaporean culture, connoisseurs focused on how the context in Singapore was different from that in other locations vis-a-vis its general treatment of migrants regardless of their origin:

> I have traveled and worked in four-five Gulf [Middle East] countries. Compared to them, the culture here is good . . . Primarily, we have a strong law and order system here. At the very least, at any time, without fearing anyone we can go to any place and come back. (Deepan)

Compared to the relatively unicultural experience of *culture campaigners*, the *cultural connoisseurs* established and maintained contact with a variety of migrant friends from different countries. The excerpt from 29-year-old Tamil-Indian worker Shekhar included in the last chapter illuminates this point. Shekhar takes active interest to reach out to other communities. When his friends were sick and were off duty, Shekhar called them and inquired about their health status. Ilangovan, another Tamil-Indian worker, noted:

> When I changed to this company, initially it was a small company . . . then it was taken over by a bigger company. So before that we have seven Singaporeans and one Indian is me. So, I am the only one to pack lunch to [. . .] the job, office.

Eventually, however, his colleagues invited Ilangovan to also eat lunch with them:

> They tell me "you come." [. . .] Yeah . . . So we go as a team outside. Every day eat different different food courts. They introduce me "this food is nice," "that food is nice." That is how . . . Yes, yes that was very good. (Ilangovan, 34)

As the above quote suggested, reaching out to the host culture was not a one-way process. Reciprocity from the host society facilitated the interaction. In the case of Ilangovan, a simple gesture like an invitation to lunch together introduced him to a new cultural world. The cultural connoisseur in him responded proactively to the situation, which endured with him throughout his, up to that point, 12-year stay in Singapore.

Just like culture campaigners, connoisseurs had a higher literacy level, and were skilled in the use of advanced mobile handsets. But unlike campaigners, this skill was not used to promote their cultural heritage. On the other hand, it was generally used to learn about other cultures. Recall the example of Victor, a 25-year-old Malayali-Indian, mentioned in chapter 6. Victor was confident of moving up in his career despite the fact that he only had a pre-university education. This he achieved by exploring the work opportunities in different countries. This orientation towards the world suggests that culture connoisseurs share characteristics with "aspirational migrants" (Burrell & Anderson, 2008) who are able to move beyond their own background and to explore the possibilities offered by globalized spaces. Connoisseurs move from region to region depending on where their career might take them. In the meantime, connoisseurs took advantage of their time in Singapore by making an effort to learn about the dominant culture there.

The Cultural Conservative

The *cultural conservatives* appeared to be somewhat biased in their preference for their own cultural heritage. This was more of an emotional and an attitudinal attachment. Rather than the promotion of their culture, as with the campaigner, or the willingness to explore alternative cultures, as with the connoisseur, the interview material shows that this group often held strong opinions about certain aspects of cultural life. This came through, for example, in views on the status of women in society and the perils of alcohol consumption. Other cultures were often evaluated based on a set of rather selective lenses. There was a more limited desire to seek out the situation of the host culture in its totality.

This perspective is seen in the situation of Ganesh, a 30-year-old Malayali-Indian. From his comments it became obvious that he was critical of what he saw as promiscuity in Singaporean culture. In comparison to his native culture, he found an unfortunate "freedom" in Singapore, which at times went against his values:

> Ours is a good joint family, life has love ["snehamulla jeevitham"] . . . relationships have great value. Here, you see, highly selfish people are in majority. Another thing. . . . In our homeland, we have mainly boys as friends. In Singapore, it's not like that . . . it's always a boy and a girl . . . full time . . . where ever they go. Everything is freely allowed. (Ganesh)

Singaporean culture was a challenge for Ganesh, as he tried to make sense of the happenings around him. Once the culture conservatives like Ganesh took exception to practices in other cultures, they often completely distanced themselves. Taking exception to these practices, they would often limit the time of their stay. This also played into their sense of belongingness to the host country. Although they spoke high of the host country on matters related to technological advancement, maintenance of law and order, and safety measures adopted at work sites, the "negative" sides of the culture outweighed the positive in most cases.

Just as they had limited interest in exploring their host culture, they had limited interest in exploring the potentials of technology. The cultural conservatives had limited technology skills. The example of Kannan and Chandran from the previous chapter help to illuminate this point. They had interaction with their own culture, but the role of communication technology was minimal in such interactions.

LINKING MOBILE PHONE APPROPRIATION TYPES WITH ACCULTURATION TYPES

So far the investigation was limited to identifying the types representing mobile phone appropriation and migrant worker acculturation. This section takes the analysis forward and links these two types. (Refer to figures 5, 6, and 7 summarizing the link between appropriation and acculturation types in appendix B).

Starting with the *culture campaigners*, the material indicates that they are likely to adopt technologies that afford group communication, since their main objective was often to quickly and efficiently link with many co-ethnics. However, the converse was not true. The appropriation type called *group communicators* was not necessarily *culture campaigners*. *Group communicators* belonged to religious and social groups and were not necessarily interested in secular expressions of culture. In such cases, *group communicators* acted merely as opinion leaders propounding religious or moral causes.

Convenience seekers and *experimenters* can be seen as *culture connoisseurs*. One type of cultural connoisseur relied on the most efficient and cost viable technological solutions to their problems. The second type of culture connoisseurs explored more on the technological possibilities, and often custom made the features of the mobile phone for their individual uses. A kind of technology-supported problem-solving attitude characterized these two types of users, which correlated with the idea of "aspirational migrant," characteristic of culture connoisseurs. The *telos* of career advancement was intricately linked to culture connoisseurs' enthusiasm in establishing friendship with co-ethnics as well as with members of other communities. Convenience seekers and experimenters as cultural connoisseurs had a world outlook rooted in pragmatism. Most of them wanted to be as neutral and secular in the multicultural spaces in which they lived and worked. Technology was utilized as a positive supplement to achieve this goal.

The *austere user* can be linked to the *culturally conservative* character. They found cultural learning as tedious as learning new mobile phone functionalities. This cultural personality type was averse both to technology appropriation at a higher level and to cultural learning. This constellation allowed little space for negotiation. Meaningful engagement with dominant host cultures was, therefore, rare.

SUMMARY

Based on the interviews with respondents, this chapter examined the quantitative results in detail. The chapter examined migrants' adaptation outcomes in the domains of work and social life and identified factors like "multiculturalism in workplace," "cultural identity," and "mobile calling to host cultures" as positively affecting the outcomes, although in irregular patterns. This suggested that among migrant workers, adaptation in one domain also influenced their adaptation to the other. The chapter also identified acculturation types and linked them to the appropriation types identified in the last chapter (Aricat et al., 2015). The final hyphenated types are:

- *convenience seeker/experimenter-culture connoisseur*
- *group communicator-culture campaigner,* and
- *austere user-cultural conservative*

The following chapter concludes the discussion by presenting theoretical implications of the findings.

NOTES

1. Descriptive statistics of variables used in the quantitative analyses is in Appendix A.

2. Cultural identity, multiculturalism in workplace, mobile phone calling to dominant host cultures, and mobile phone calling to home culture.

3. Adaptation to social life and organizational commitment.

4. Demographics was the focus of the first block. It explained 5.8% of the variance in adaptation to social life, $F(4, 435) = 6.7, p < .001$. Age ($\beta = .16$, $p < .01$) had a stronger effect compared to education ($\beta = .14$, $p < .01$).

5. The second block with acculturation factors explained 1.8% of the variance in the dependent variable, $F(2, 433) = 4.18, p < .05$. Multiculturalism in workplace ($\beta = .13, p < .05$) was the only significant predictor.

6. The third block that considered mobile communication explained 0.8% of the variance in adaptation to social life, $F(2, 431) = 1.83, p > .05$.

7. The fourth block with interaction terms explained 0.5% of the variance in the dependent variable, $F(4, 427) = .63, p > .05$.

8. Demographics, entered as the first block, explained 1.8% of the variance in organizational commitment, $F(4, 435) = 2.02, p > .05$.

9. The second block of variables describing acculturation explained 37.4% of the variance, $F(2, 433) = 128, p < .001$. Cultural identity ($\beta = .33, p < .001$) had a slightly higher effect than multiculturalism in workplace ($\beta = .32, p < .001$).

10. The third set of variables examining mobile communication explained 2.4% of the variance, $F(2, 431) = 8.54, p < .001$, with mobile phone calling to dominant host cultures significantly affecting organizational commitment ($\beta = .17, p < .001$).

11. The interaction terms were entered as the fourth block, $F(4, 427) = 5.01, p < .01$. Interaction between multiculturalism in workplace and calling to host culture was significant at $p < .01$ level ($\beta = .14$). The model that included interaction terms predicted 2.7% of variance.

12. Refer to table 9 in appendix A for interviewee details.

13. Grid is a brand name of push-to-talk phones widely used among Singapore's industrial workforce. The website says that the technology was developed by Motorola, and it "provides its users the benefits of a trunked radio and a cellular telephone." The added advantage to the workforce is that this technology "places more users in a given spectral space, compared to analog cellular and two-way radio systems, by

using speech compression and time division multiple accesses (TDMA)." See https://www.grid.net.sg/service/iden/

14. Caste is a feature of South Asian societies, especially in the Hindu tradition. Occupational closure based on one's lineage and endogamy within one caste group are its chief characteristics. Those groups who traditionally did scavenging, tanning, and butchery were considered outcaste and were systematically discriminated against by the upper caste groups.

15. Solomon (2016) notes in relation to gradual disappearance of lower caste identities of Tamil population in Singapore: "The growth of a diasporic consciousness and the concomitant broadening of notions of identity away from immediate kin, village and caste groups greatly opened the space for imaginative modes of resistance against caste prejudice" (p. 99).

16. This is a kind of surveillance that crosses boundaries and follows a migrant to her destination. Fellow villagers report any kind of "misdemeanor" of the immigrant back to the elders in the village. "Transnational village gaze" ensures that the immigrant is not deviating from the village norms after moving to the city.

17. A business and meeting place in Little India, where South Asians congregate on weekends.

18. As mentioned in the introduction (chapter 1), these are the divisions on which the immigration policy of Singapore government is based. Foreign talent is entitled to various benefits in Singapore including opportunities for naturalization and family visa status. Workers, on the other hand are employed on short-term contracts and are considered guests.

19. When trade and settlement started under the British in the early nineteenth century, most of the low-skilled labor force was recruited from the southeastern coast of India. This Tamil-speaking community later settled in today's Malaysia and Singapore, eventually making their culture a part of contemporary Singaporean culture.

Chapter Eight

Conclusion

In this volume we have attempted to understand the role of mobile communication among low-skilled male transient workers from South Asia as they seek their position in Singapore, and as they work to maintain contacts with their home culture. Two theoretical frames—one related to technology appropriation and the other related to acculturation—were followed in the book. The model of technology appropriation helped us understand phenomena like the functional and instrumental uses of the phone and the symbolic meanings derived by the users. The bi-dimensional acculturation model helped us theorize the connections established and maintained by the migrants along the axes of retention of cultural identity, and fostering of relationships with people from other cultures.

An Asian setting—Singapore—was chosen for the study, which emerged as an attractive destination of low-skilled migrant workers from South and Southeast Asia and China in the recent decades. We elaborated on the context in which male Bangladeshi and Indian migrant workers immigrated to Singapore for work. Most of them wanted to leverage on the wage differentials between the sending and receiving countries, as they decided to migrate. Unlike the so-called foreign talent (typically professional workers), these migrant workers were not provided opportunities for naturalization. Living singly in the host country, this section of workers felt the greatest need to retain communication links with their left behind families in the country of origin, as the vignette of Francis in the beginning of this book showed. In addition, they also had to acculturate to the host society, which in the Singaporean context constituted a number of ethnicities. Migrants' acculturation needs were particularly observed in the domains of work and social life. This book examined the role of mobile technology in their intercultural engagement in these two domains.

Although this research began with the objective of understanding the impact of mobile phone on migrant acculturation, over the course of the study it was observed that the technology, along with having an impact over the users from a macro perspective, was appropriated in different ways as part of social processes. Meanings were ascribed to mobile phone affordances based on the cultural and social background of the users, while users appropriated its functionalities as determined by the social context (Hall, 1980; Jamison & Hard, 2003). The findings fall in line with the social constructivist position, namely that the interpretive frames within which a technology is adopted and used strongly frames the technology (MacKenzie & Wajcman, 1999). Without dismissing the "impact" and "effects" of mobile phone communication that the populations world over have witnessed from a macro perspective, the research calls for a deeper engagement with the interpretive schemas adopted by users as they appropriate the technology. Similarly, the essentialist nature of typological thinking in acculturation has also been critiqued in this research with the introduction of acculturation types that are more dynamic and fluid.

The key findings of the study are:

- among the transient migrant workers, mobile phones help retain ties with those left behind;
- mobile phones help coordinate work in multicultural teams of low-skilled workers;
- the mutual help and understanding developed in multicultural teams extend to social spaces, where friendships are developed across cultural groups;
- there are instances when mobile communication does not help overcome barriers within organization, especially when workers communicate with authorities who are higher up in hierarchy; and
- there are instances when migrants' own cultural norms pose barrier to forging closer ties with other cultures.

To conclude the discussion, we would like to structure the contributions of this work under three broad themes: communication, cultural adaptation, and power relations. This helps locate our study in the larger fields of sociology and intercultural communication.

COMMUNICATION

In line with earlier studies, our investigation revealed that the male transient migrants had the communication needs. They needed to retain links with families left behind as well as friends and colleagues in the host country (Diminescu, 2008; Metykova, 2010). The findings also brought to fore several nuances in this trend. To recapitulate, most immigrants considered family communication as indispensable. Mobile phones helped ensure their "connected presence" with family in such cases (Licoppe, 2004). Migrants' extant links with family were also strengthened by mobile phones as gifts. In work spaces, mutual help, coordination, and problem solving remained central for migrant workers.

The study also brought to fore several nuances in migrants' communication with home. For example, family communication was not always acceptable to them during work and most of them resented the demand for gifts from community members back home. These nuances in migrants' home communication, and their engagement with the home culture has implications for theory. For instance, contrary to the conception of the bi-dimensional model, migrants' engagement with home culture is neither simple nor linear in nature. It is irregular and non-linear, involving complex negotiations. Migrants do not relate to their home culture in a unilateral way, but co-construct the meaning of their culture with others. Migrants follow performative routines in this co-construction of meaning. This problematizes the way in which culture is theorized as monolithic in the bi-dimensional model. The hypothetical question raised by the bi-dimensional model, that is, "whether you would like to retain your cultural identity," thus needs revision. Following the findings of this study, the question can be revised as "whether you would like to retain your cultural identity (and how), given that you are a migrant and you have access to a range of communication technologies." This nuanced understanding of migrants' engagement with home culture, situated in their use of mobile technology, is one of the contributions of this study.

Further, the nuances in migrants' engagement with home shows cultural identity in a new light. The findings do not contest the point that transient workers have nationalistic feelings, and they often want to retain their cultural heritage. On the other hand, the findings highlight the need to consider the layered processes involved in one's identification with one's culture. Mediation by mobile technologies is a crucial aspect that was delineated in this process. Depending on the utilization and appropriation of mobile technologies, "group communicators," say, differed markedly from "austere users" in their identification with and performativity of their home culture. This led us to follow a two-pronged approach and delineate the hyphenated categories of appropriation-acculturation among immigrants. Theoretically, this suggests a revision on identifying immigrants as part of a "national container society" (Wimmer & Glick Schiller, 2002, p. 324). Although belonging to the same nation or culture, the way the immigrants identify with these ideas differs substantially.

In sum, communication is central to immigrants' identification with their culture. This identification, however, is not a unilinear process, but involves several complexities. Mobile communication allows for a negotiation in a mediated space, which in turn affects the meaning of culture and cultural identity for immigrants.

CULTURAL ADAPTATION

The findings from this work suggest that the question of cultural adaptation of labor migrants needs to be rethought. For low skilled migrant workers, the adaptation challenges are mainly related to work and social life. Mutual help, coordination, and problem solving remain central to their adaptation in organizations. The friendships developed in the organizational spaces are carried over to the social spaces. Membership in co-ethnic networks is instrumental in receiving help and achieving coordination. Further complicating this picture, the findings suggested that the immigrants followed a subjective interpretation of adaptation. Mobile technology skills, aspirations, and socio-economic status were important when immigrants subjectively defined their adaptation and well-being in the country.

At Work

Adaptation to a multicultural organization was a prerequisite for migrants' survival and success in the host society. Two types of immigrants were found to be committed to their work organizations: first, workers who had high identification with their home culture; and, two, workers who had an open disposition to other cultures in workspaces. These effects were mediated by mobile communication, which played a major role in achieving coordination, problem solving, and mutual help within co-ethnic and multicultural groups. This eventually led to immigrants' enhanced commitment to the organization. These findings have theoretical implications, as discussed below.

Prior research, for example in the similarity-attraction paradigm, suggested that people felt favorably disposed to those who were similar to them in a group. This proclivity towards similarity had positive effects including greater communication within group and enhanced group effectiveness (Chatman et al., 1998; Tsui et al., 1992). Conversely, members of minority cultures in an organization were likely to perceive threat to their identities, which caused negative attitudes toward others. This has been associated with negative outcomes like low organizational commitment (Riordan & Shore, 1997; Rupert et al., 2010). Although literature suggested significant relationships between group identification and organizational commitment, the processes underlying these relationships were largely unknown. The findings of this study addressed this gap by explaining the underlying processes in the light of mobile communication. The study found that when co-ethnic immigrants came together in an organization and identified themselves with a work group, they achieved better coordination. In culturally homogeneous work groups, the migrant workers leveraged on their national, linguistic, and cultural identities to gain better acceptance, ensure information access, and seek professional assistance in times of need. Thus, the findings advance knowledge by highlighting how mobile communication is crucial while explaining the organizational commitment of this category of workers.

The findings also suggested that the migrants who were favorably disposed to colleagues from other cultures showed higher levels of organizational commitment. This supported the evidence available from prior research. Workers who were open to diversity in organization in terms of race, culture, and gender enjoyed positive outcomes like cognitive complexity, adaptability, integration, and learning (Chae & Foley, 2010; Ely & Thomas, 2001; Frankenberg et al., 2013; Riordan & Shore, 1997; Rupert et al., 2010; Tadmor, et al., 2009). In our study, migrants gained a footing in the new culture by engaging with a multicultural work team. This interaction was often via the mobile phone, in the context of work, as they jointly completed the organizational tasks. While language was a barrier in such communication, they eventually learned the working class patois of English (or a third language like Hindi, in the case of South Asian workers) from their colleagues. Interaction over mobile phone mainly involved micro-coordination related to logistic matters (Ling & Yttri, 2002). Despite the brevity of the communication and the instrumental nature of the interaction, there was always a learning process involved, which ultimately helped their adaptation.

Demographics did not decide migrants' organizational commitment. This implies that in a multicultural work context, acculturation-communication factors have greater power to predict workers' organizational commitment, in comparison to their demographics. Positive outcomes in organizations may not be achieved only by ensuring that the workforce is young, is well paid and well educated. On the contrary, issues related to cultural identity and multicultural disposition of the workforce need increased attention. This finding has particular relevance to the Singaporean situation, where demographics are unduly emphasized while allowing immigrants to the country. Based on the findings, we suggest that relevant policy measures are adopted to enhance the multicultural attitude of the migrant workers. This helps the integration of minority members into the organization and, ultimately to the host society as a whole.

The model that predicted organizational commitment was statistically powerful within the limits of social science research. The model explained 43% variance in the outcome variable. It suggests that organizations would benefit if they make a periodic assessment of cultural identity, mobile calling behavior, and multiculturalism of the migrant workers at the lower stratum. The leap in the predictive power of the statistical model with the introduction of acculturation variables (R^2 value increased by 37%) suggests that the two factors—cultural identity and multiculturalism in workplace—are decisive in understanding low-skilled migrants' commitment to the organization.

Organizational commitment has an extended impact on other favorable outcomes including productivity and job satisfaction. It is, thus, important to understand what factors influence migrant workers' organizational commitment. Studies have shown that committed employees show greater loyalty to the organization, and are more satisfied in their work in comparison to employees who are less committed (Mathieu & Zajac, 1990; Riketta & Van Dick, 2005). As an important predictor of job related outcomes, organizational commitment needs greater emphasis in statistical models that studies acculturation-communication of migrant workers.

In Social Spaces

A large part of the general population's knowledge about the male transient workers is based on how they are seen in the social spaces. The portrayal of immigrants in media and on the internet pertains to their engagement with the resident civil society. This is somewhat different for the female transient workers who often worked in the homes of the resident population (Thomas & Lim, 2011). Thus, male immigrants' interactions in social spaces are more easily adjudged as good or bad. In fact, stereotypes are easily formed (e.g., some immigrants as unruly and uncivilized) when interactions go wrong in social spaces. These negative effects are mitigated for immigrants and their adaption is usually facilitated by their friendship networks. Co-ethnic networks, sometimes formed in the work spaces, and extended to social spaces, help the male immigrants to face adaptation challenges in their social life. This blurring of boundaries between the professional and social is a major theme in mobile communication literature (Mante-Meijer & van de Loo, 1998).

Blurring of boundaries between the social and the professional is more relevant in the context of developing economies, where the low-income population juggles between income-generating as well as purely expressive and socially oriented tasks. The informal character of their professions and their reliance on social networks in day to day business results in this crossing of boundaries between the social and the professional very often. The reasons for the blurring of social and professional domains for the migrant workers are rooted in the challenges they face in cultural adaptation. A notable example of this is the peaceful co-existence of different ethnic groups in around 50 purpose-built dormitories and innumerable worksite accommodations in Singapore. Although the concepts used by scholars vary from friendships, conviviality to multiculturalism, the basic process involves cultural adaptation beyond the confines of the organization into social spaces. This study helped advance knowledge in the area by extending the "blurring" concept to the intercultural context, thereby identifying the challenges faced by the immigrants in day-to-day life and their strategies to cope with them.

The findings also support the idea of "integration" as the most effective strategy of acculturation. According to the studies that have followed the bidimensional model, immigrants who relate to other cultures without severing ties with their own culture eventually have an easier time adapting to the host society (e.g., Horenczyk & Munayer, 2007; Krishnan & Berry, 1992; Piontkowski et al., 2000). This study has shown that this remained to be true not only for immigrants with greater permanency but also for transient workers of low economic standing.

Adaptation to social life has a meaning for transient labor migrants, which is different from that of the professional or business class, who have opportunities for naturalization. For the labor migrants, adaptation operated at a more fundamental level—at the level of basic survival. The basic requirements for these workers included learning the language, feeling comfortable in the new country, and learning to abide by the rules of the host country. The findings suggested that those who had an open disposition towards colleagues from other countries could learn the survival strategy better than those who were closed towards colleagues from different cultures. It can be seen in the examples of Ilangovan, Palanivel, and Shekhar noted above (see chapters 6 and 7). Palanivel and Shekhar developed friendships with colleagues from Thailand and Myanmar and followed up with these outside work. We can also see it in that Ilangovan's colleagues from the local Singaporean population introduced him to local foods and tastes, which facilitated his adaptation to social life. Indeed, the reciprocity of the host society is a crucial factor that decides to what extent transient workers can follow their strategy of "integration."

POWER RELATIONS

Immigrants constantly encountered power relations both in their communication and in their cultural adaptation. These power relations ranged from the soft control exerted by the family on the immigrant to the more subtle and internalized control exerted by social norms on the immigrants. Migrants' interaction with the management was often characterized by direct and explicit display of power relations.

There were migrants who strategized their communication with home, by for example discouraging family communication during work. Migrants also resented the soft control exerted by family or by the community. This control included demands for gifts for community members and also the more indirect disciplinary gaze enforced by village norms. Some immigrants willingly accepted the norms embedded in their home culture, as they aligned their mobile and social media activities in line with these. The example of Victor (see chapter 7), who reiterated the village norms in his interactions with the host society, substantiated this argument.

Power relations regularly figured in immigrants' interaction with employers and work. Vineeth's (see chapter 7) example of mobile communication not helping resolve issues with management substantiates this point. The prohibition on mobile phone in the name of following safety practices illuminates power relations. The discourse of "safety" is appropriated by the authority to discipline the workers at the lower stratum. Senior level workers like Pranab and Deepan subscribed to this discourse, while the material presented here shows that many workers resented it.

In addition to identifying power as exerted by any individual or emanating from a source, we have examined it in the day-to-day interactions. The top-down nature of power has been widely characterized (e.g., in the form of state control, host society's demands, and organization's impositions). Power acts in an extraneous fashion in such cases. In this study, we also considered it as intrinsic to the migrants' interactions with the home culture. Indeed, the material shows that power relations emerged and manifested themselves when migrants interacted with the home culture (see chapter 6). This is also seen in the kind of self-disciplining migrants learn in the host society as the meritocratic ideology takes precedence. It is also seen in the way they define their pleasures and aspirations. In other words, the power relations play a major role in constituting the subjectivity and identity of immigrants.

An important contribution of this research is that the layered meanings and implications of acculturation have been made relevant to migrants at the lower strata. It highlighted the disadvantaged position of low-skilled migrants in their negotiation for freedom and rights in the host country. The economic logic of labor migration itself presents an unstable acculturation context for the low-skilled migrants and the host society. Due to this same economic logic, a management perspective has taken precedence over a political praxis of migrant rights and assertive action. Management practices are rarely sympathetic to the complex processes by which the cultural adaption occurs among the labor migrants. An economistic perspective has taken precedence in the policy circles on low-skilled labor migration the world over. Our attempt in this book was to understand the complex processes of acculturation by keeping individual migrants at the center of analysis.

LIMITATIONS OF THE STUDY

As with any academic work, there are limitations to the analysis done here. First, due to the convenience sampling methods used in the study, the extrapolations based on statistical results has limitations. Second, the population studied had specific demographic characteristics—male low-skilled migrant workers from four linguistic groups from Bangladesh and India. Hence, the results of the study may have only limited generalizability to other groups of migrant workers such as foreign domestic workers, professionals, and immigrant entrepreneurs. Third, the mobile communication measures used in the quantitative part of the study have taken into account only one aspect, namely, mobile voice phone calling. The effects of other modes including Web 2.0 features like microblogging, social networking, and instant chatting may also have a bearing on migrant acculturation. The research design did not take this into account, particularly because it was only starting to enter the lives of the migrants at the time of data collection. Fourth, the questionnaire survey relied on self-report and each respondent had to fill in the questionnaire by himself, albeit assisted by the researcher or their friends in some cases. Due to insufficient skills in filling up a questionnaire, the respondents with lower education level might not have provided information true to their best knowledge and belief. Since the survey used self-reported measures, all limitations associated with this method were applicable to this study too.

Fifth, the interviews were conducted in only three languages—English, Hindi, and Malayalam. Although the services of interpreters were sought whenever necessary, some information might have lost when the respondent expressed it in his mother tongue and the researcher understood it in another language. Sixth, the researcher did not have access to any of the living places of migrants, usually dormitories, and to their work sites. Dormitory owners denied permission to the researcher to access these facilities. Had the researcher gained access to these sites, some information provided by the respondents could have been verified and cross-checked. In the absence of it, respondents' verbal recollections and narratives had to be taken to be true. Seventh, acculturation is a bi-dimensional process for both migrants as well as the host society. It is only through an investigation involving both migrants and the host society that a vivid picture of this phenomenon would emerge.

This study, instead, looked only at the experience of migrants and highlighted their experience over that of the host society. Eighth, on thorough scrutiny there might appear a disconnect between the conceptual definition and the operationalization of the variable "mobile communication to dominant host cultures." While the six questions in this scale (three each for duration and frequency of calling) enquired on the amount of communication to friends from other cultures and to colleagues, it is to be noted that the latter category might have some overlaps with the mobile communication measure on home culture calling. However, the differential effect of mobile calling to home and host cultures on the adaptation outcomes shows that such overlaps have been minimal and insignificant.

Ninth, the data collection process lasted for more than two years, during which period many factors related to migrant workers and mobile technologies could have changed. Although a descriptive understanding of such changes was achieved by qualitative interviews, the questionnaire might have evoked different meanings for respondents who filled it early on and those who filled it towards the end of the data collection period. For example, voice over internet calling diffused to the migrant population at a fast pace during the data collection period. The question "for how long do you talk to family back in your home country" might have evoked different connotations for respondents who filled the questionnaire before and after the smartphone diffusion. Tenth, in the regression analyses, the block with mobile communication variables as well as the full model of "adaptation to social life" has low explanatory power (R-square value). Hence, there is limited scope for generalizing the results.

Finally, the research followed a positivist approach as it investigated how the mobile phone was appropriated within a cross-cultural context. Drawing on mainly from the tradition of Enlightenment rationality and the sociological thinking emerged from that event/period, this study failed to fully appreciate latest understanding related to the de-centerdness of meaning in a text (Hall, 1992). The conceptualization and execution of the study, interpretation of the phenomenon of technology appropriation and the critique of existing research literature would have been different had the study drew more from the philosophical tradition that negated the originality of the knowing subject.

FUTURE RESEARCH

The quantitative part of the study assessed the relationship between acculturation, mobile communication and adaptation. Future research can build sophisticated models and causality structures that include these factors. Mobile communication should eventually be introduced into the acculturation models, complementing the interpersonal and mass media communication variables used in these models.

The concept and design followed by this research may be extended to include other migrant segments in Singapore, but also international and in-country migrant populations around the world. With newer affordances of smartphones, and the fast diffusion of social networking Apps, it has become essential for acculturation research to investigate thoroughly the social appropriation of these affordances among immigrants. New situated stories on immigrant acculturation can be identified and narrated if future research can focus on these affordances and their appropriation by immigrants. It is relevant in the emerging migrant context to understand how individuals and groups appropriate mobile technology to empower themselves in non-institutionalized ways. Findings from such studies can guide the society to go beyond official narratives and media apprehensions on the cultural threats posed by migrants and understand the newer patterns of cultural mixing and plurality with better clarity.

There should be sustained research interest on the acculturation experience of migrants at the lower strata, which eventually unfolds in diverse patterns in disparate demographic groups. Indeed, the ultimate challenge is to acknowledge that the acculturation experience of each individual migrant is irreducible and unique.

APPENDIXES

APPENDIX A: TABLES

Table 1. Bangladeshi migrants: Labor migration from Bangladesh between 2001 and 2010 (Category-wise).

	Worker's category				
Year	*Professional*	*Skilled*	*Semi-skilled*	*Less-skilled*	*Total*
2001	5,940	42,742	30,702	109,581	188,965
2002	14,450	56,265	36,025	118,516	225,256
2003	15,862	74,530	29,236	134,562	254,190
2004	12,202	110,177	28,327	122,252	272,958
2005	1,945	113,655	24,546	112,556	252,702
2006	925	115,468	33,965	231,158	381,516
2007	676	165,338	183,673	482,922	832,609
2008	1,864	292,364	132,825	448,002	875,055
2009	1,426	134,265	84,517	255,070	475,278
2010	387	90,621	20,016	279,678	390,702

Source: Bureau of Manpower Employment and Training (http://www.bmet.gov.bd/BMET/resources/Static%20PDF%20and%20DOC/publication/Migration-BOEP.pdf)

Table 2. Descriptive: Description of the six variables used in the regression analyses. For details on the items included in each variable, refer to tables on Factor Analyes.

	N	*Minimum*	*Maximum*	*Mean*	*SD*
Multiculturalism in workplace	440	1	5	4.098	.93
Cultural identity	440	1	5	4.38	.81
Amount of calling to home	440	.63	5	3.048	.71
Amount of calling to host	440	0	5	2.618	.95
Adaptation to social life	440	1.93	7	5.569	.69
Organizational commitment	440	1.21	5	3.969	.8

Source: Created by the authors.

Table 3. Acculturation items: Results of factor analysis on acculturation items adapted from MIRIPS.

Factors and items (eigenvalues in parentheses)	*Mean*	*SD*	*Factors[1]* 1	2	3	4	5	6	7	8	9	10
Cultural identity (6.44)												
I think of myself as [ethnic].	4.29	1.26	**.767**	.032	.089	.007	.003	.181	–.042	–.148	.014	.044
I feel that I am part of [ethnic] culture.	4.17	1.26	**.747**	.149	.112	–.055	–.011	.045	.03	–.113	.012	–.004
I am proud of being [ethnic].	4.56	1.01	**.825**	–.047	.094	–.042	.076	.063	–.055	.139	.024	.067
I am happy to be [ethnic].	4.51	1.04	**.767**	–.014	.072	–.05	.049	.12	–.102	.162	.03	–.055
Being part of [ethnic] culture makes me feel happy.	4.46	1.04	**.731**	–.027	.121	–.127	.033	.081	–.057	.158	.05	.013
Being [ethnic] makes me feel good.	4.33	1.16	**.594**	.045	.165	–.094	–.028	.065	.149	.207	.033	–.079
Being part of [ethnic] culture is embarrassing to me. (Reverse)	4.31	1.27	**.418**	–.117	.168	**–.538**	.018	–.076	–.039	.184	.064	.063
Being [ethnic] is uncomfortable for me. (Reverse)	4.4	1.13	**.411**	–.164	.103	**–.559**	.025	–.035	–.039	.079	–.009	.217
Cronbach's alpha			*.85*									
% variance explained			*11.77*									
Factor excluded from later analysis (3.86)												
I prefer social activities which involve [ethnic group] members only.	3.26	1.58	**.273**	**.628**	–.077	.098	–.011	–.025	–.06	–.187	–.017	.11

(continued)

Table 3. ***(continued)***

Factors and items (eigenvalues in parentheses)	*Mean*	*SD*	*Factors*[1] *1*	*2*	*3*	*4*	*5*	*6*	*7*	*8*	*9*	*10*
I prefer social activities which involve Singaporeans only.	2.92	1.5	.028	**.705**	.039	.043	.17	.172	–.057	.14	–.034	–.109
I prefer to have only Singaporean local friends.	2.67	1.56	–.103	**.686**	.027	.146	.086	.182	–.056	.021	.066	.022
I don't want to have either Singaporean or [ethnic] friends.	2.04	1.4	–.238	**.49**	.012	**.413**	.035	**–.275**	–.043	.114	–.107	–.03
I prefer to have only [ethnic] friends.	2.45	1.5	.021	**.618**	–.064	**.289**	–.137	–.134	–.044	–.116	.13	.07
Cronbach's alpha				*.72*								
% variance explained				*7.3*								
Multiculturalism in workplace (2.69)												
It is a pleasure to work with people of different ethnicities.	4.31	1.14	.161	.062	**.75**	–.118	.124	.107	–.011	.172	.002	.061
An organization that has a variety of ethnic and cultural groups is more able to tackle new problems as they occur.	3.82	1.33	.03	–.093	**.683**	–.05	.222	–.004	–.149	.185	.096	–.078
It is best for an organization if all workers forget their different ethnic and cultural differences and work in unison.	4.35	1.15	**.251**	.087	**.698**	–.139	.007	.076	.085	.078	–.106	.033

We should recognize that cultural and racial diversity is a fundamental characteristic of my organization.	3.91	1.34	.207	–.101	**<u>.691</u>**	–.003	.076	.182	.013	–.189	–.038	.036
Cronbach's alpha					*.74*							
% variance explained					*6.7*							
Factor excluded from later analysis (1.61)												
I feel that [ethnic group] should maintain our own cultural traditions and not adapt to those of Singapore.	2.64	1.51	.088	.247	.021	**<u>.402</u>**	**<u>–.524</u>**	–.067	–.118	.008	.011	–.037
I don't want to attend either Singaporean or [ethnic] social activities.	2.01	1.35	–.048	.241	–.224	**<u>.56</u>**	.001	–.133	–.065	.144	–.041	.202
I feel that [ethnic group] should adopt the Singaporean cultural traditions and not maintain those of our own.	2.17	1.41	–.082	.135	–.06	**<u>.463</u>**	**<u>.408</u>**	–.123	.087	–.001	.125	–.073
I am concerned about losing my cultural identity.	2.18	1.41	.009	.044	–.001	**<u>.622</u>**	–.087	–.029	–.13	.014	.117	.166
Cronbach's alpha						*.6*						
% variance explained						*6.64*						
Factor excluded from later analysis (1.45)												
Even if I lose my mother tongue eventually, I should be more competent to use languages used in Singapore.	2.6	1.51	.099	.076	.095	.185	**<u>.402</u>**	–.18	**<u>–.252</u>**	–.023	**<u>–.334</u>**	.202

(continued)

Table 3. *(continued)*

Factors and items (eigenvalues in parentheses)	*Mean*	*SD*	*Factors[1]* 1	2	3	4	5	6	7	8	9	10
I think I have become a part of Singaporean society.	3.29	1.4	.005	.19	.198	.058	**.691**	–.02	–.03	.104	.019	.061
I will be happy to adopt Singaporean culture.	3.45	1.4	.091	.054	.158	–.095	**.754**	.196	.065	.045	.068	–.024
It is more important to me to be fluent in languages in Singapore than in [ethnic language].	2.78	1.51	.108	.117	.034	**.367**	**.44**	.157	–.033	–.15	**–.349**	–.016
Cronbach's alpha							*.61*					
% variance explained							*6.17*					
Factor excluded from later analysis (1.31)												
It is important for me to be fluent in both [ethnic language] as well as the languages used in Singapore.	3.83	1.32	**.298**	–.141	.04	.046	.134	**.53**	–.208	–.121	**.254**	–.014
I feel that [ethnic group] should maintain our own cultural traditions but also adopt those of Singaporean.	3.86	1.3	.208	.032	.101	–.069	.034	**.596**	.134	.244	–.005	–.02
I prefer social activities that involve both Singaporeans and [ethnic] members.	3.67	1.41	.065	**.399**	.215	–.123	.135	**.603**	–.1	.052	–.005	.031

I prefer to have both [ethnic] and friends from dominant communities in Singapore.	4.19	1.18	.201	.081	**.261**	–.106	.01	**.511**	.039	.193	**–.295**	.245
Cronbach's alpha								*.58*				
% variance explained								*4.75*				
Factor excluded from later analysis (1.16)												
An organization that has a variety of ethnic or cultural groups has more problems with productivity than organizations with one or two basic cultural groups. (Reverse)	2.75	1.47	–.079	–.237	–.009	.02	.052	.033	**.746**	–.221	.064	–.108
The productivity of the organization is undermined by workers of different ethnic and cultural backgrounds sticking to their old ways. (Reverse)	3.1	1.49	–.014	–.003	–.034	–.144	.062	–.066	**.738**	.220	–.058	.113
Cronbach's alpha									*.46*			
% variance explained									*3.91*			
Factor excluded from later analysis (1.15)												
I have more than 5 close friends from the local communities.	3.84	1.34	.193	–.091	**.26**	.015	**.275**	.154	.076	**.471**	–.124	–.085
In Singapore, I have more than 5 friends from my own ethnic community.	4.05	1.3	**.279**	–.011	.124	–.01	–.028	.17	–.007	**.659**	.078	–.043

(continued)

Table 3. ***(continued)***

			Factors[1]									
Factors and items (eigenvalues in parentheses)	*Mean*	*SD*	*1*	*2*	*3*	*4*	*5*	*6*	*7*	*8*	*9*	*10*
Cronbach's alpha										*.52*		
% variance explained										*3.91*		
Factor excluded from later analysis (1.1)												
It is more important to me to be fluent in [ethnic]/Hindi languages than in languages used in Singapore.	3.23	1.46	.205	.119	.002	.156	.032	.031	–.016	.006	**<u>.791</u>**	.084
Cronbach's alpha											*N/A*	
% variance explained											*3.26*	
Factor excluded from later analysis (1.05)												
It is not important for me to be fluent either in [ethnic language] or language used in Singapore.	2.69	1.49	–.007	.028	.038	.098	.069	.074	.018	–.068	.061	**<u>.88</u>**
Cronbach's alpha												*N/A*
% variance explained												*3.05*

1. Factor loadings above 0.25 (both negative and positive dimensions) have been highlighted. While grouping the items and identifying the factors, the highest factor loading in an expected direction has been given priority.

Source: Created by the authors.

Table 4. Socio cultural adaptation items: Factor analysis results for adaptation to social life (questions adapted from MIRIPS "sociocultural competence" scale)

			Factors				
Factors and items (with eigenvalues in parentheses)	*Mean*	*SD*	*1*	*2*	*3*	*4*	*5*
Complying with the host society rules (4.48)							
Obeying traffic rules as a pedestrian	6.15	1.22	**.728**	.104	–.004	.198	–.181
Obeying rules in public transport while traveling	6.06	1.27	**.726**	.068	–.017	**.303**	–.143
Obeying anti-littering law	5.89	1.49	**.738**	.086	.142	–.096	.156
Obeying anti-drug law	5.86	1.56	**.679**	.114	.079	–.09	.244
Overall order and discipline in society	6.05	1.3	**.552**	.12	.08	.044	**.470**
Cronbach's alpha			*.75*				
% variance explained			*13.11*				
Competency in host society language(s) (1.93)							
Dealing with people in authority	5.67	1.23	.024	**.595**	.144	.033	–.074
Communicating with people of a different ethnic group	5.46	1.23	.124	**.659**	.062	.158	.041
Communicating in a language other than your mother tongue	5.2	1.38	.155	**.719**	.043	.135	.044
Learning a language to know about the culture it represents	5.26	1.47	.023	**.614**	.03	–.008	.249
Cronbach's alpha				*.62*			
% variance explained				*11.13*			
Adaptation to host society's custom and lifestyle (1.49)							
Family relationships in host society	4.74	1.46	–.019	.047	**.53**	**.302**	**.365**
Metro culture	5.37	1.49	.081	.054	**.428**	**.408**	.204

(continued)

Table 4. *(continued)*

			Factors				
Factors and items (with eigenvalues in parentheses)	*Mean*	*SD*	*1*	*2*	*3*	*4*	*5*
Dress code of women	3.88	1.71	.071	.106	**.791**	–.077	–.106
Male-female relationships	4.31	1.65	.017	.07	**.758**	.093	.011
Overall world view of this society	5.23	1.34	.166	.129	**.429**	**.229**	**.260**
Cronbach's alpha					*.67*		
% variance explained					*10.31*		
Factor excluded from later analysis (1.1)							
Festivities of groups other than yours	5.44	1.32	.03	.139	–.004	**.691**	.099
Nightlife	5.27	1.57	.039	.032	**.273**	**.613**	–.031
Cronbach's alpha						*.38*	
% variance explained						*9.24*	
Factor excluded from later analysis (1.04)							
Status given to workers in this society	5.2	1.41	.136	**.225**	.178	**.252**	**.613**
Cronbach's alpha							*N/A*
% variance explained							*6.4*

Source: Created by the authors.

Table 5. Organizational commitment items: Factor analysis results for organizational commitment

			Factors					
Factors and items (with eigenvalues in parentheses)	*Mean*	*SD*	*1*	*2*	*3*	*4*	*5*	*6*
Factor 1 (8.96)								
I am willing to put in a great deal of effort beyond that normally expected in order to help this organization be successful.	4.24	1.2	**.627**	**.45**	–.146	–.029	–.06	–.073
I talk up this organization to my friends as a great organization to work for.	4.12	1.2	**.71**	**.226**	.198	–.004	–.057	–.07
I would accept almost any type of job assignment in order to keep working for this organization. (Item excluded from later analysis)	3.87	1.37	**.450**	**.311**	.076	–.188	–.191	**–.329**
I find that my values and the organization's values are very similar.	3.82	1.27	**.569**	.134	**.386**	.175	.035	–.143
I am proud to tell others that I am part of this organization.	4.07	1.22	**.732**	.116	.196	.151	.023	–.105
This organization really inspires the very best in me in the way of job performance.	4	1.24	**.635**	**.259**	**.351**	–.026	.054	–.04
I am extremely glad that I chose this organization to work for over others I was considering at the time I joined.	4	1.23	**.61**	.212	.229	.003	.088	–.082
Deciding to work for this organization was a definite mistake on my part (Reverse). (Item excluded from later analysis)	3.84	1.37	**.485**	.042	.098	**–.310**	**.345**	.158
I feel a sense of pride in working for this organization.	4.03	1.23	**.693**	.235	**.366**	.033	.051	.097
This organization has a tradition of worthwhile accomplishments.	3.86	1.24	**.483**	.227	**.427**	.067	.129	.078

(continued)

Table 5. ***(continued)***

Factors and items (with eigenvalues in parentheses)	*Mean*	*SD*	*Factors* 1	2	3	4	5	6
Cronbach's alpha			*.88*					
% variance explained			*16.48*					
Factor 2 (2.86)								
I really feel as if the organization's problems are my problems.	3.94	1.22	**.389**	**.493**	**.376**	.037	.036	.077
The most important things that happen to me involve my work.	3.91	1.26	**.228**	**.502**	**.428**	–.019	–.11	–.105
I live, eat and breathe my job.	3.85	1.34	.058	**.57**	**.383**	–.103	–.04	**–.378**
I complete my assigned duties on time.	4.31	1.06	.22	**.668**	.221	.043	–.041	0
I comply with the rules and regulations of this organization.	4.49	0.97	**.509**	**.664**	–.011	.04	–.011	.022
I do what my boss says without complaint.	4.26	1.12	.166	**.651**	.096	.199	–.006	–.044
At times I wonder how I have begun to feel more committed towards my work after coming to Singapore.	4.09	1.17	**.322**	**.47**	.236	.134	.129	–.043
Cronbach's alpha				*.82*				
% variance explained				*10.69*				
Factor 3 (1.58)								
For me this is the best of all possible organizations for which to work.	3.63	1.35	**.353**	.171	**.605**	.033	.043	–.157
I have warm feelings toward this organization as a place to work live and work with.	3.89	1.22	**.459**	**.347**	**.509**	.015	.083	.01

I would be quite willing to spend the rest of my career with this organization.	3.22	1.5	.213	.228	**<u>.677</u>**	.044	–.027	.114
Since joining this organization, my personal values and those of the organization have become more similar.	3.7	1.2	**<u>.383</u>**	.075	**<u>.611</u>**	.204	.059	–.061
Cronbach's alpha					*.77*			
% variance explained					*9.81*			
Factor excluded from later analysis (1.29)								
I really care about the fate of this organization.	3.55	1.42	.208	.211	.037	**<u>.600</u>**	–.086	.123
I volunteer for tasks that are not required of me.	3.19	1.46	.027	.273	–.007	**<u>.468</u>**	–.241	–.021
In my job here, I sometimes have to act in ways that are not completely consistent with my true values.	3.09	1.42	–.054	–.042	–.009	**<u>.699</u>**	–.176	–.055
I attend functions that are not required, but that help the organization's image.	2.84	1.51	–.029	–.065	**<u>.276</u>**	**<u>.714</u>**	–.038	–.136
Cronbach's alpha						*.63*		
% variance explained						*7.16*		
Factor excluded from later analysis (1.1)								
I feel very little loyalty to this organization. (Reverse)	3.32	1.65	.047	.006	**<u>–.299</u>**	**<u>.315</u>**	**<u>.672</u>**	.088
It would take very little change in my present circumstances to cause me to leave this organization. (Reverse)	2.66	1.37	–.019	–.075	–.017	–.159	**<u>.450</u>**	.198
There's not too much to be gained by sticking with this organization indefinitely. (Reverse)	3.55	1.39	**<u>.413</u>**	–.067	.236	–.246	**<u>.443</u>**	.014

(continued)

Table 5. *(continued)*

Factors and items (with eigenvalues in parentheses)	*Mean*	*SD*	*Factors*					
			1	*2*	*3*	*4*	*5*	*6*
Unless I'm rewarded for it in some way, I see no reason to expend extra effort on behalf of this organization. (Reverse)	2.99	1.48	–.053	.105	.131	–.174	**.585**	–.068
Cronbach's alpha							*.41*	
% variance explained							*6.09*	
Factor excluded from later analysis (1.0)								
I could just as well be working for a different organization as long as the type of work was similar.	2.46	1.48	–.112	.062	.006	–.1	.06	**.845**
Cronbach's alpha								*N/A*
% variance explained								*3.92*

Source: Created by the authors.

Table 6. Mobile communication items: Factor analysis results for items on amount of calling on mobile phone to home and host cultures.

Factors and items (eigenvalues in parentheses)	*Mean*	*SD*	*Factors* 1	2	3
Factor 1 *(4.01)*					
On average, for how long do you talk to your family over phone at a time?	3.44	1.18	**.542**	–.117	**.434**
For how long do you talk to your friends back home at a time over phone?	2.7	1.22	**.704**	–.028	.219
For how long do you talk to your friends in other countries over phone at a time?	2.06	1.34	**.676**	.174	–.067
For how long do you talk to your friends from your own community in Singapore?	2.46	1.34	**.666**	.021	.190
For how long do you talk to your friends in other communities in Singapore?	1.82	1.27	**.734**	.315	–.099
For how long do you talk to your fellow-workmen, at or below your level?	2.06	1.26	**.672**	.234	–.093
For how long do you talk to someone above your status in office over phone?	1.74	1.28	**.598**	**.405**	–.270
Cronbach's alpha			*.8*		
% of variance explained			*22.84*		
Factor 2 *(1.86)*					
How often do you call your friends in other countries?	2.18	1.57	**.323**	**.514**	.001
How often do you call your friends from other communities in Singapore?	2.94	1.74	.132	**.732**	.175
How often do you call your fellow-workmen, at or below your level?	3.87	1.42	.017	**.646**	.232

(continued)

Table 6. ***(continued)***

			Factors		
Factors and items (eigenvalues in parentheses)	*Mean*	*SD*	*1*	*2*	*3*
How often do you call anyone above your status in office?	3.27	1.73	.112	**.695**	.004
Cronbach's alpha				*.65*	
% of variance explained				*16.55*	
Factor 3 *(1.37)*					
How often do you call your family back home?	4.51	0.95	–.065	–.040	**.729**
How often do you call your friends back home?	3.16	1.31	.145	.310	**.601**
How often do you call your friends from your community in Singapore?	3.87	1.37	–.009	.394	**.617**
Cronbach's alpha					*.55*
% of variance explained					*12.33*

Source: Created by the authors.

Table 7. Demographic variables

N=464	*Range*	*Mean*	*SD*
Age	21–50	29.01	5.51
Education	5 to 21.5 years	12.09	2.22
Duration of residence	1 month to 14 years	49.71 months	34.95
Ethnicity			
Malayali: 74[1] (15.9%)			
Tamil: 175 (37.7%)			
Telugu: 102 (22%)			
Bengali: 113 (24.4%)			
Income			
Below SGD 500: 45 (9.7%)			
Between SGD 500 and SGD 999: 312 (67.2%)			
Between SGD 1,000 and SGD 1,999: 96 (20.7%)			
SGD 2,000 or above: 11 (2.4%)			

1. To compensate for Malayali's lower presence in the questionnaire sample, interview sample included more of them: 46.7%. The remaining respondents in the interview sample were evenly distributed among the three other ethnicities.

Source: Created by the authors.

Table 8. Hierarchical regression analyses: Hierarchical regression analyses (to test hypotheses 1 and 2) of the impact of cultural identity, multiculturalism in workplace, mobile phone calling to home and host cultures on two adaptation outcomes—adaptation to social life and organizational commitment.

Predictors	*Adaptation to social life (H1),* β	*Organizational Commitment (H2),* β	*Δ R^2 Adaptation to social life*	*Δ R^2 Org. Commitment*
Block 1: Demographics				
Age	**.157****	.043		
Education	**.142****	–.05		
Period of residence	–.026	.024		
Income	.085	–.045	.058	.018
Block 2: Identity and multiculturalism				
Cultural identity	.028	**.333*****		
Multiculturalism in workplace	**.134***	**.321*****	.018	.365
Block 3: Mobile phone calling				
Mobile calling to home culture	.058	–.002		
Mobile calling to dominant host cultures	.045	**.17*****	.008	.024

(continued)

Table 8. *(continued)*

Predictors	*Adaptation to social life (H1), β*	*Organizational Commitment (H2), β*	*Δ R^2 Adaptation to social life*	*Δ R^2 Org. Commitment*
Block 4: Interaction terms				
Cultural Identity X Calling to home culture	.081	–.068		
Cultural Identity X Calling to host cultures	–.051	.095		
Multiculturalism X Calling to home culture	–.075	.016		
Multiculturalism X Calling to host cultures	.012	**.136****	.005	.027
R Square			.089	.433
Adjusted R Square			.063	.417

Notes: ***p < .001; **p < .01; *p < .05

Source: Created by the authors.

Table 9. Details of interviewees

Sl. no.	*Interviewee ID*[1] *(Ethnicity)*	*Age*	*Education (In years; including vocational training)*	*Income slabs*[2]	*Job*	*Duration of residence in Singapore (In months; under different contracts)*	*Interview date and venue*[3]
1	BGL01 Mahim	29	10.0	2	Mechanic, Shipyard	43	21-Jul-11; Penjuru[4]
2	BGL02 Jaffer	24	12.0	2	Steel reinforcement (skilled worker), Construction	24	14-Jun-11; Soon Lee Rd.
3	BGL03 Siddiqui	24	9.0	1	General worker, Construction	24	13-Jun-11; Soon Lee Rd.
4	BGL04 Niyaz	30	10.0	3	Welder, Construction	72	5-Jul-11; Soon Lee Rd.
5	BGL05 Faizal	31	15.0	2	General worker, Construction	12	17-Jul-11; Penjuru
6	BGL06 Kishore	30	10.0	2	General worker, Construction	12	10-Aug-11; Penjuru
7	BGL07 Deepak	23	12.0	1	General worker, Construction	25	30-Jul-11; Penjuru
8	BGL08 Kabir	25	10.0	2	General worker, Construction	18	3-Jul-11; Soon Lee Rd.
9	BGL09 Prabodh	25	12.0	1	General worker, Construction	36	10-Jun-11; Soon Lee Rd.
10	BGL10 Shariful	23	10.0	2	General worker, Process Construction & Maintenance (PCM)	30	3-Jul-11; Soon Lee Rd.
11	BGL11 Ismail	37	11.0	2	Fitter, Shipyard	18	3-Aug-11; Soon Lee Rd.

(continued)

Table 9. *(continued)*

Sl. no.	*Interviewee ID*[1] *(Ethnicity)*	*Age*	*Education (In years; including vocational training)*	*Income slabs*[2]	*Job*	*Duration of residence in Singapore (In months; under different contracts)*	*Interview date and venue*[3]
12	BGL12 Sultan	21	12.0	2	Rigger signalman, Construction	24	14-Jun-11; Soon Lee Rd.
13	BGL13 Niranjan	24	10.0	2	Air-con mechanic, Construction	24	14-Jun-11; Soon Lee Rd.
14	BGL14 Khurram	32	12.0	3	Safety supervisor, Construction	168	15-Jun-11; Soon Lee Rd.
15	BGL15 Abdullah	27	10.0	2	Welder, Shipyard	84	21-Jul-11; Penjuru
16	BGL16 Ilyas	29	12.0	2	Security guard, Construction site	60	25-Aug-11; Soon Lee Rd.
17	BGL17 Pranab	40	12.0	2	Supervisor, Construction	120	15-Jun-11; Soon Lee Rd.
18	BGL18 Wazim	32	8.00	1	General worker, Construction	36	9-Jun-11; Soon Lee Rd.
19	BGL19 Pervez	28	15.0	2	Painter, Construction	36	8-Jun-11; Soon Lee Rd.
20	BGL20 Salim	26	10.0	2	Safety supervisor, Construction	44	19-Jul-11; Penjuru
21	BGL21 Nawaz	31	15.5	3	Skilled worker, Construction	117	21-Sep-12; Centurion
22	BGL22 Gautam	34	16.0	4	Office work, Construction	22	09-Mar-13; Near Teka Mall
23	MAL01 Lawrence	26	13.0	3	Welder, Refinery	60	31-Jul-11; Penjuru

24	MAL02 James	36	13.0	2	Quality controller, Marine sector	60	27-May-11; Eng Lee hostel
25	MAL03 Kesavan	35	12.0	3	Quality controller, PCM	72	30-May-11; Eng Lee hostel
26	MAL04 Dileep	42	10.0	2	Insulator, PCM	12	16-Jul-11; Penjuru
27	MAL05 Deepan	40	10.0	2	Air-con mechanic, Shipyard	48	17-Jul-11; Penjuru
28	MAL06 Chandran	45	10.0	2	General worker, Shipyard	11	3-Apr-11; Penjuru
29	MAL07 Bhaskaran	50	12.0	2	Pipe fitter, Shipyard	84	21-Jul-11; Penjuru
30	MAL08 Andrews	27	12.0	2	Quality controller, PCM	84	27-May-11; Kian Teck hostel
31	MAL09 Ganesh	30	10.0	2	Welder, Shipyard	72	31-Jul-11; Penjuru
32	MAL10 Varghese	32	10.0	2	Electrician, Shipyard	33	4-Jun-11; Soon Lee Rd.
33	MAL11 Benny	25	12.0	2	Quality controller, PCM	18	3-Jun-11; Blk 916
34	MAL12 Murukan	32	14.0	2	Electrician, Shipyard	36	4-Jun-11; Soon Lee Rd.
35	MAL13 Viswanath	26	12.0	3	Electrician, Shipyard	42	4-Jun-11; Soon Lee Rd.
36	MAL14 Bhuvan	34	12.0	2	Technician, PCM	24	2-Jun-11; Soon Lee Rd.
37	MAL15 Francis	30	12.0	2	Quality analyst, PCM	84	11-Aug-11; Avery Lodge
38	MAL16 Victor	25	12.0	2	Welder, PCM	60	25-May-11; Blk 988C

(continued)

Table 9. *(continued)*

Sl. no.	*Interviewee ID*[1] *(Ethnicity)*	*Age*	*Education (In years; including vocational training)*	*Income slabs*[2]	*Job*	*Duration of residence in Singapore (In months; under different contracts)*	*Interview date and venue*[3]
39	MAL17 Shaju	31	12.0	2	Quality analyst, Piping sector	60	17-Jul-11; Penjuru
40	MAL18 Suthan	33	12.0	3	Quality controller, Welding	60	17-Jul-11; Penjuru
41	MAL19 Kannan	23	10.0	2	General worker, Insulation	14	19-Jul-12; Penjuru
42	MAL20 Jayan	33	12.0	2	General worker, Insulation	26	24-Mar-11; Soon Lee Rd.
43	MAL21 Cleetus	41	10.0	2	General worker, Construction	21	16-Jul-11; Penjuru
44	MAL22 Mahesh	25	12.0	1	General worker, Insulation	24	3-Apr-11; Penjuru
45	MAL23 Narendran	28	10.0	2	General worker, PCM	23	24-Mar-11; Soon Lee Rd.
46	MAL24 Vineeth	23	14.0	1	General worker, Shipyard	8	8-Apr-11; Soon Lee Rd.
47	MAL25 Kiran	30	14.0	2	Electrician, PCM	24	23-May-11; Eng Lee hostel
48	MAL26 Jayesh	29	14.0	3	Quality controller, Welding	24	6-Jun-11; Blk 816
49	MAL27 Sudheesh	31	10.0	2	Sheet material fitter, Insulation	30	18-Aug-12; Joo Koon
50	MAL28 Roopesh	28	11.0	2	Welding, PCM	72	18-Aug-12; Joo Koon

51	MAL29 Pavithran	42	13.0	2	Cladding & Insulation	36	18-Aug-12; Joo Koon
52	MAL30 Naveen	32	13.0	3	Marine engine mechanic	84	18-Aug-12; Joo Koon
53	MAL31 Janardanan	24	14.0	3	Piping supervisor	18	19-Aug-12; Joo Koon
54	MAL32 Manoj	23	12.0	2	Forklift driver, Shipyard	12	1-Sep-12; Joo Koon
55	MAL33 Thampi	30	16.0	3	Mechanical supervisor, Construction	84	9-Jun-13; Bluestar
56	MAL34 Peter	30	16.25	2	Fitter, Refinery	44	9-Jun-13; Eng Lee hostel
57	MAL Cristo	35	14	4	Quality controller–Welding	42	15-Mar-2013; Blk 843
58	TAM01 Afsal	43	12.0	3	Quality controller, Refinery	48	23-Jul-11; Penjuru
59	TAM02 Palanivel	26	12.0	2	General worker, Insulation	72	23-Jul-11; Penjuru
60	TAM03 Kathiresan	27	16.0	2	Quality controller, PCM	36	3-Aug-11; Soon Lee Rd.
61	TAM04 Hussain	25	10.0	2	Safety supervisor, Oil & Petrochemical	60	21-Aug-11; Penjuru
62	TAM05 Shekhar	29	14.0	2	Fitter & Rigger signalman	36	22-Aug-11; Penjuru
63	TAM06 Ilangovan	34	10.0	2	Skilled worker, Insulation	62	30-Apr-11; Soon Lee Rd.
64	TAM07 Marimuthu	26	15.0	2	General worker, Insulation	30	18-Aug-12; Joo Koon
65	TAM08 Prakash	25	15.0	2	General worker, Insulation	30	18-Aug-12; Joo Koon

(continued)

Table 9. *(continued)*

Sl. no.	*Interviewee ID[1] (Ethnicity)*	*Age*	*Education (In years; including vocational training)*	*Income slabs[2]*	*Job*	*Duration of residence in Singapore (In months; under different contracts)*	*Interview date and venue[3]*
66	TAM09 Ilavarasan	28	12.0	2	General worker, Insulation	27	19-Aug-12; Joo Koon
67	TAM10 Kaviraj	36	10.0	2	Store keeper, PCM	36	25-Aug-12; Joo Koon
68	TAM11 Daniel	38	12.0	2	Store keeper, Construction	60	25-Aug-12; Joo Koon
69	TAM12 Venkatesh	24	16.0	2	Office work, Oil & Petrochemical	72	9-Mar-13; Penjuru
70	TAM17 Selvan	49	10	4	Driver	216	02-Feb-13: Blk 843
71	TEL01 Paarthiban	22	11.0	2	Welder, Shipyard	36	25-Jul-11; Blue Star
72	TEL02 Balakrishna	25	11.0	2	Welder, Construction	48	16-Jul-11; Penjuru
73	TEL03 Siddharth	25	16.0	2	Store keeper, Construction	13	19-Aug-12; Joo Koon
74	TEL04 Karthik	29	10.0	2	Fitter & Welder, Construction	72	24-Feb-13; Soon Lee Rd.
75	TEL05 Ramesh	28	15.0	2	Painting & Plastering, Construction	60	24-Feb-13; Soon Lee Rd.

1. BGL for Bangladeshi or Indian Bengali, MAL for Malayali, TAM for Tamil, and TEL for Telugu.
2. Slab 1, below SGD 500; Slab 2, between SGD 500 and 999; Slab 3, between SGD 1,000 and 1,999; Slab 4, SGD 2,000 and above. Income is calculated inclusive of pay for overtime work.
3. Although the dormitory names are mentioned in this column, the actual venue of the interview was in a park or secluded area near to dormitory.
4. Centurion and Bluestar are dormitory names, whereas Joo Koon, Penjuru, and Soon Lee Rd. are dormitories identified with place or road name.

Source: Created by the authors.

Table 10. Themes from interview data: List of mobile appropriation and acculturation categories and different migrant types' extent of fit to each category.

	Typologies		
Mobile appropriation categories	*Culture connoisseurs/ Convenience seeker, Experimenter*	*Cultural conservative—Austere user*	*Culture campaigner—Group communicator*
Functional use[1]	Frequent; Both feature phones and smartphones are used.	Frequent, but with careful consideration of cost. Low diffusion of smartphones.	Frequent; Low-cost messages widely used to reach a group of people.
Symbolic use[1]	Consider mobiles as status symbols	Gift mobiles to loved ones	No symbolic use observed
Use of advanced features[1]	Frequent	Seldom	Seldom
Cost considerations related to mobiles usage[2]	Average	High	Average
Content of meta-communication[2]	Frequent deliberations on what is new in the mobiles market	On a look-out for uncluttered information in mediated spaces, but also from friends	Frequent deliberations on the low-cost methods available on mobiles to communicate with group members
Acculturation categories			
Amount of communication to home country[1]	High, but independent of the communication to host cultures	High, at the expense of sustained engagement with the host cultures	High; News from home country strengthen cultural identity
Amount of communication to members of dominant host cultures[1]	High, irrespective of high amount of communication to home	Low; Engagements restricted to professional spaces and to close friends from other cultures	Low

(continued)

Table 10. *(continued)*

	Typologies		
Acculturation categories	*Culture connoisseurs/ Convenience seeker, Experimenter*	*Cultural conservative—Austere user*	*Culture campaigner—Group communicator*
Extent of relationship with dominant host country cultures[1]	High	Low	Low
Identification with one's own culture[1]	Low	Average	Very high
Level of engagement with co-ethnics in host country[1]	Average	Average	Very high

1. Pre-determined categories.
2. Categories emerged with data analysis.

Source: Created by the authors.

APPENDIX B: FIGURES

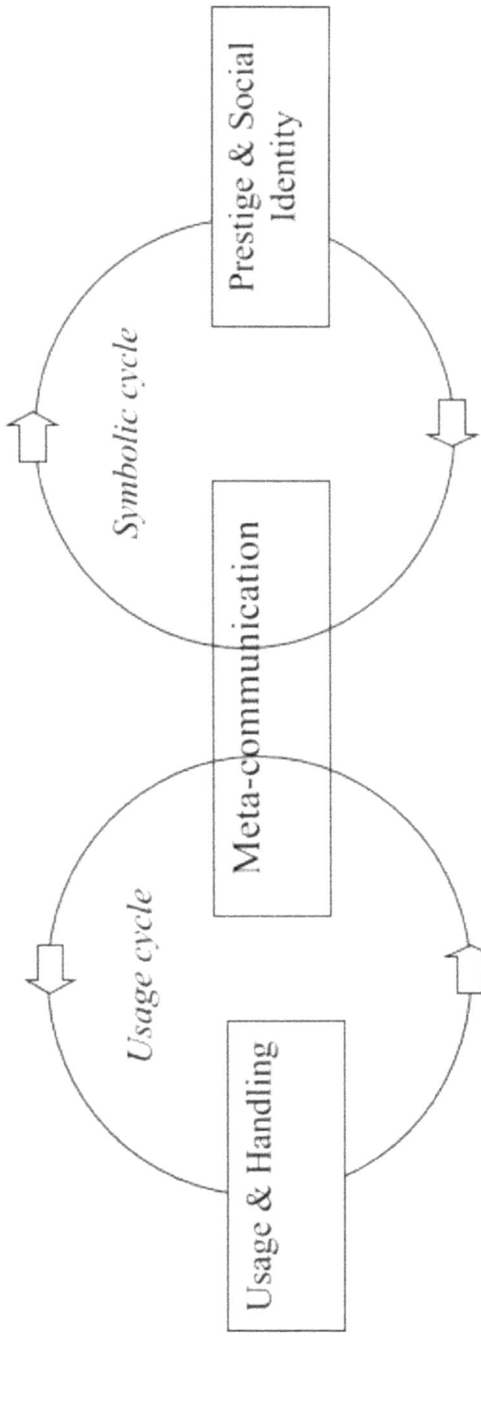

Figure 1. The circular model of mobile phone appropriation.
Source: Karnowski, von Pape, & Wirth, 2008

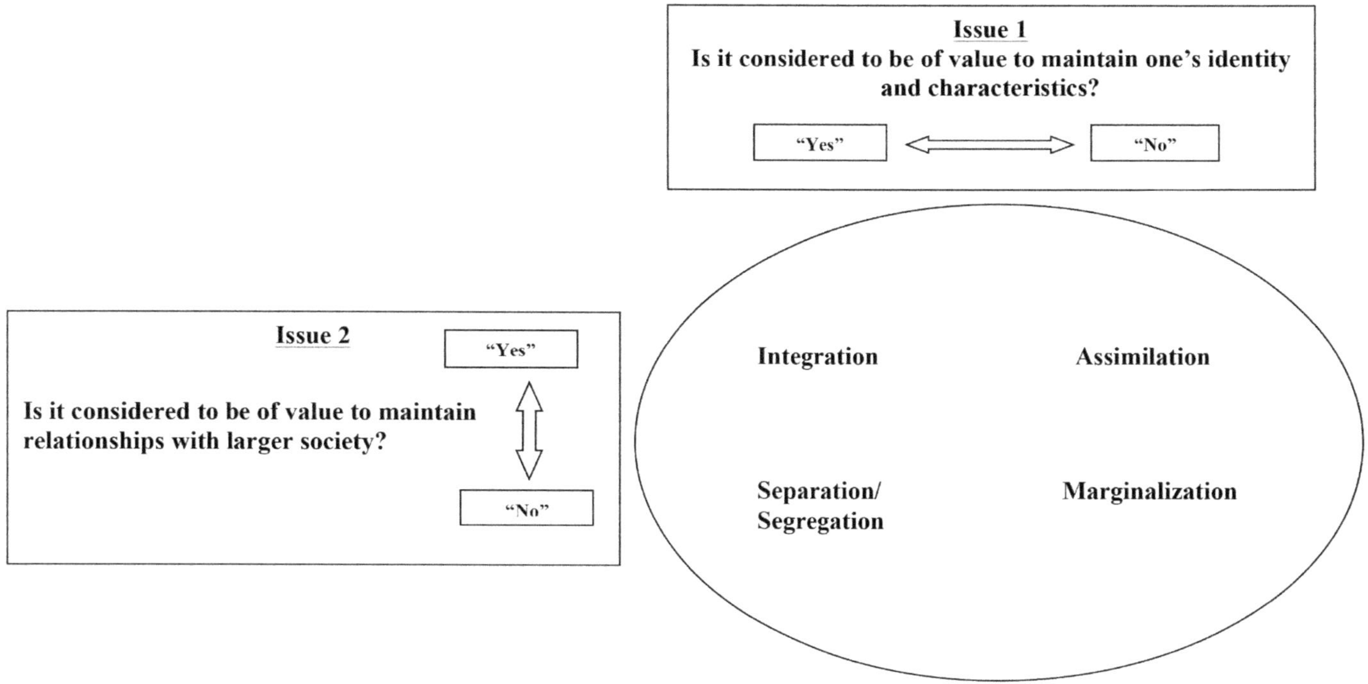

Figure 2. Acculturation orientations.
Source: Berry, 1997

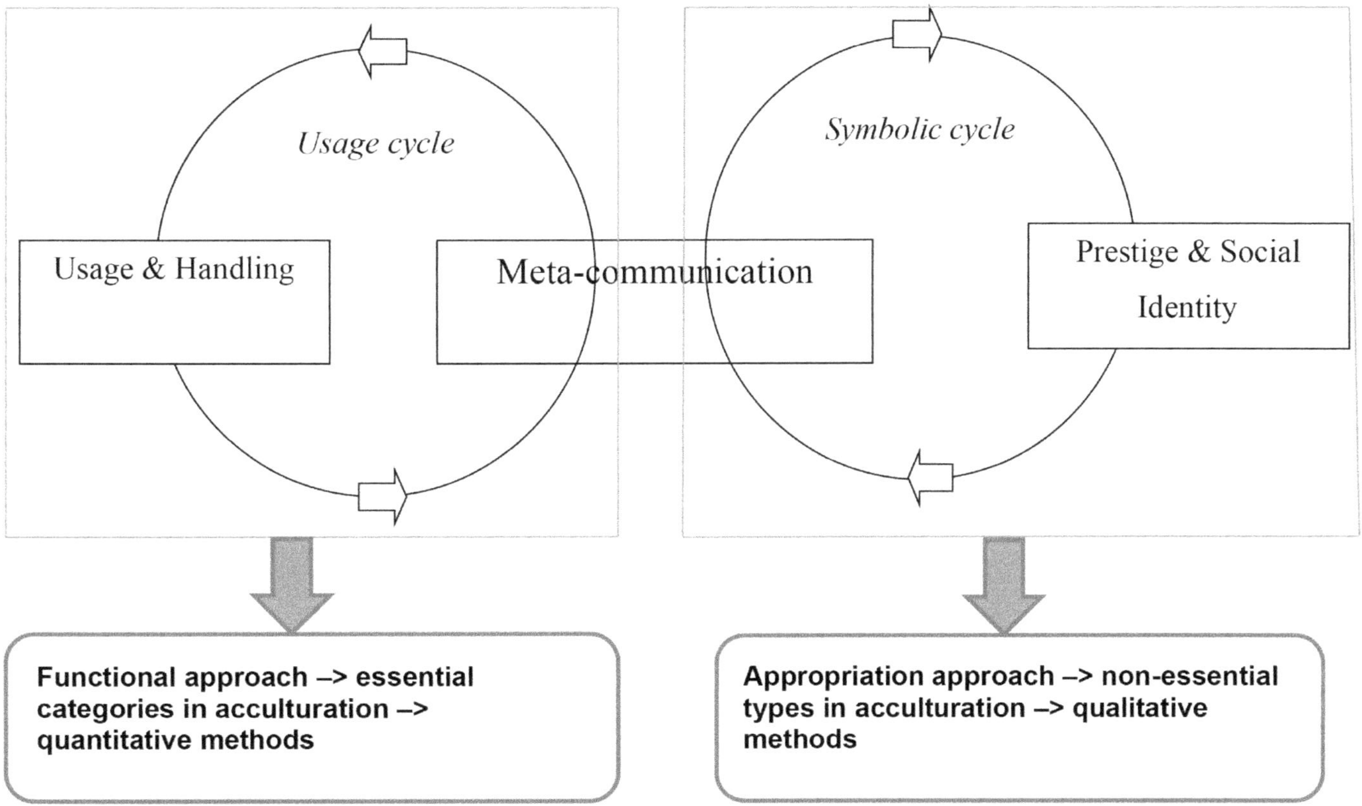

Figure 3. Integrated diagram.
Source: Created by the authors.

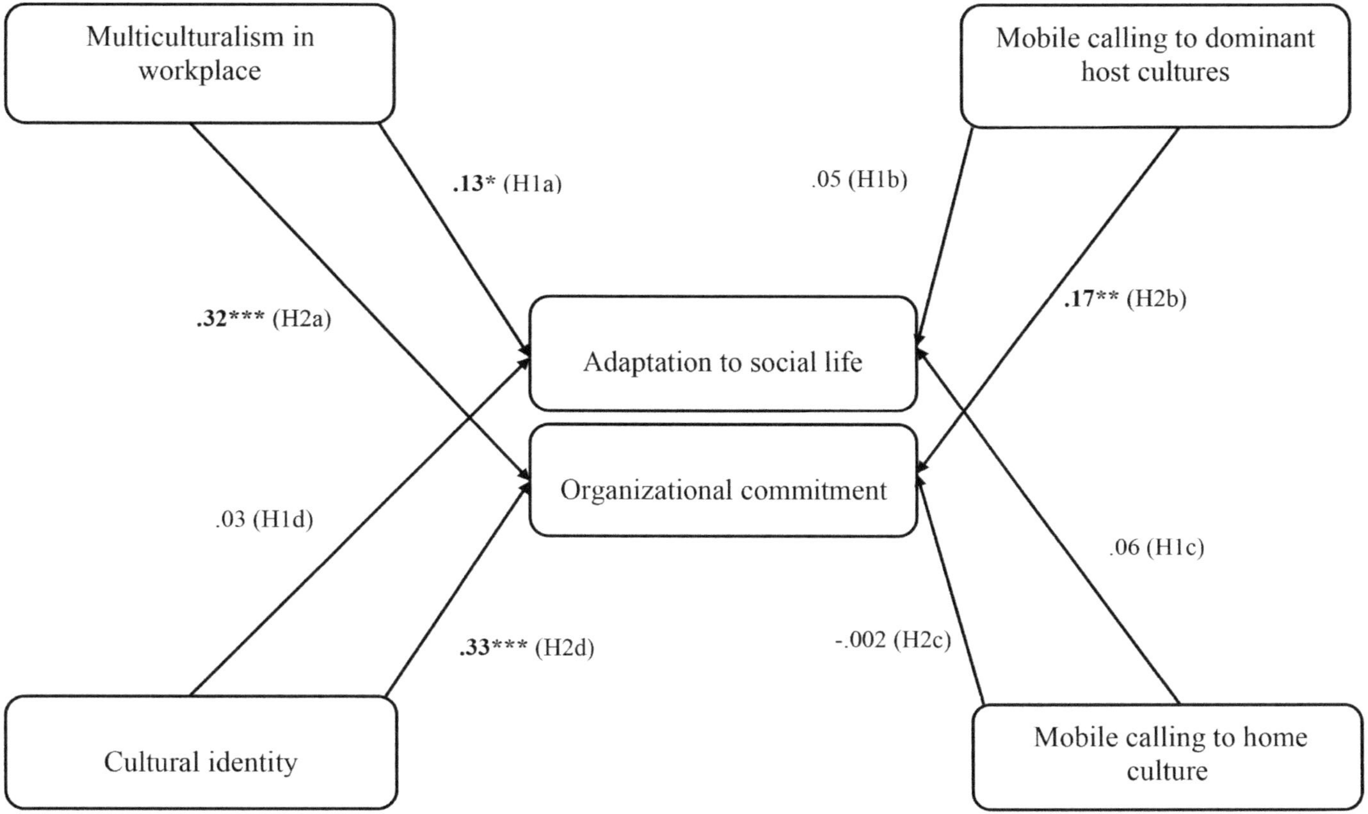

Figure 4. Beta on model: The model diagram with standardized beta coefficients corresponding to each hypothesis.
Source: Created by the authors.

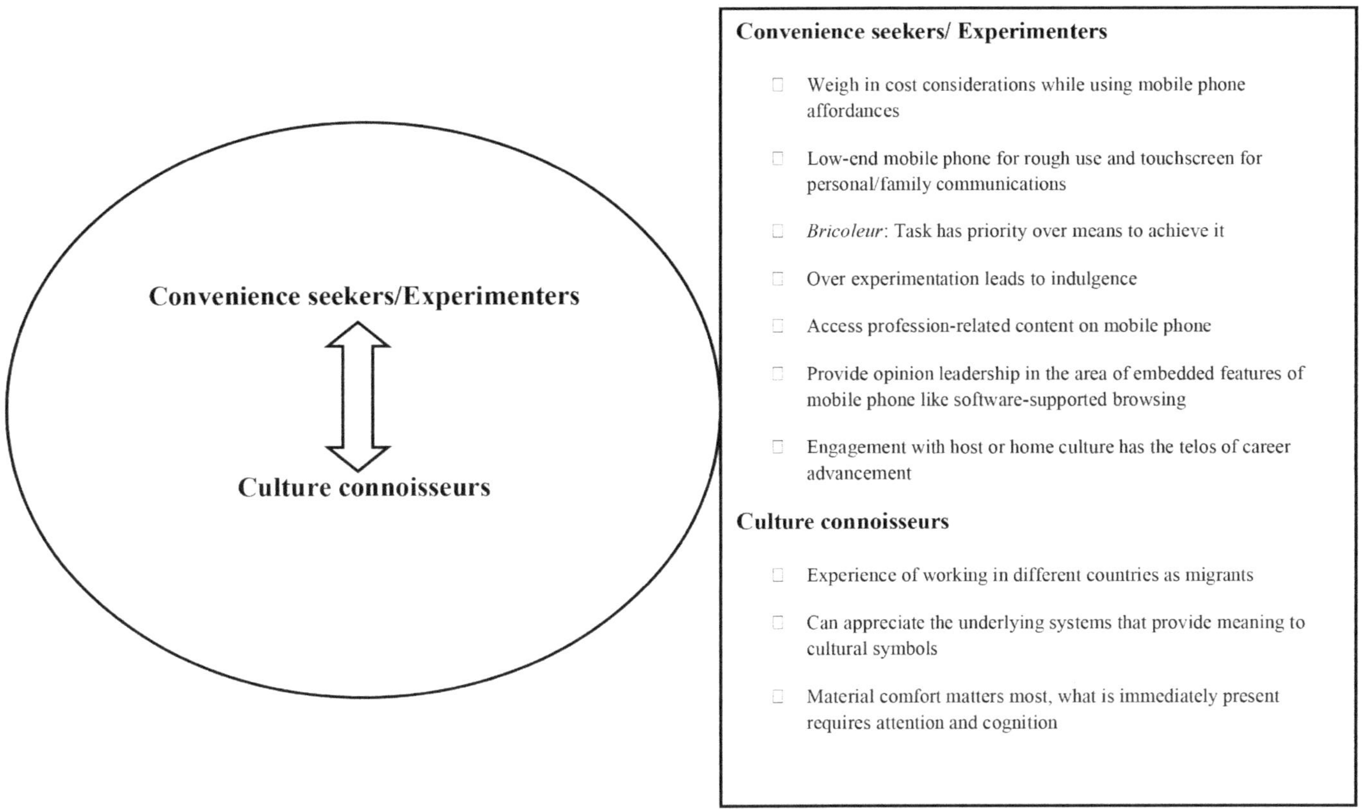

Figure 5. Diagram summarizing the main characteristics of Convenience seekers/experimenters-Culture connoisseur.

Source: Created by the authors.

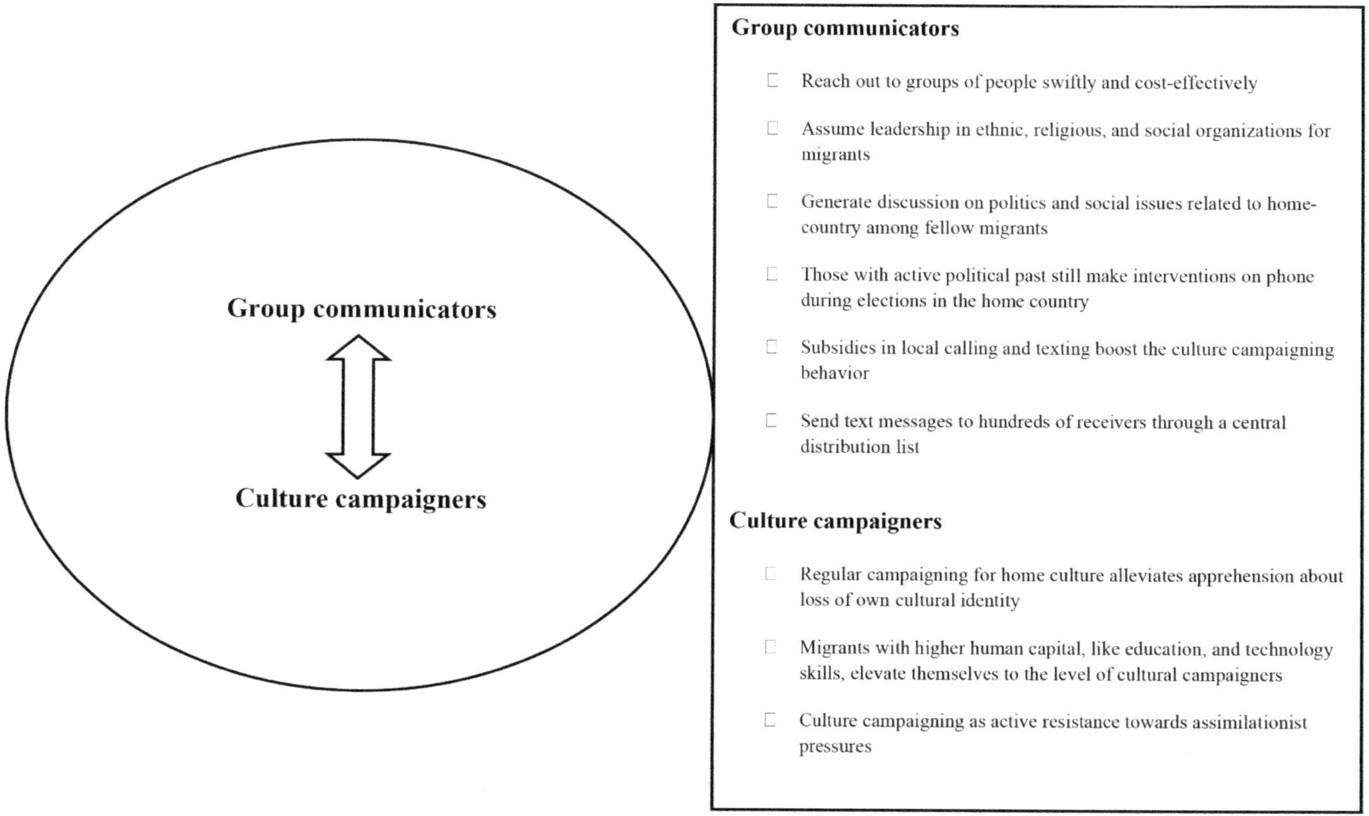

Figure 6. Diagram summarizing the main characteristics of group communicators-culture campaigners.

Source: Created by the authors.

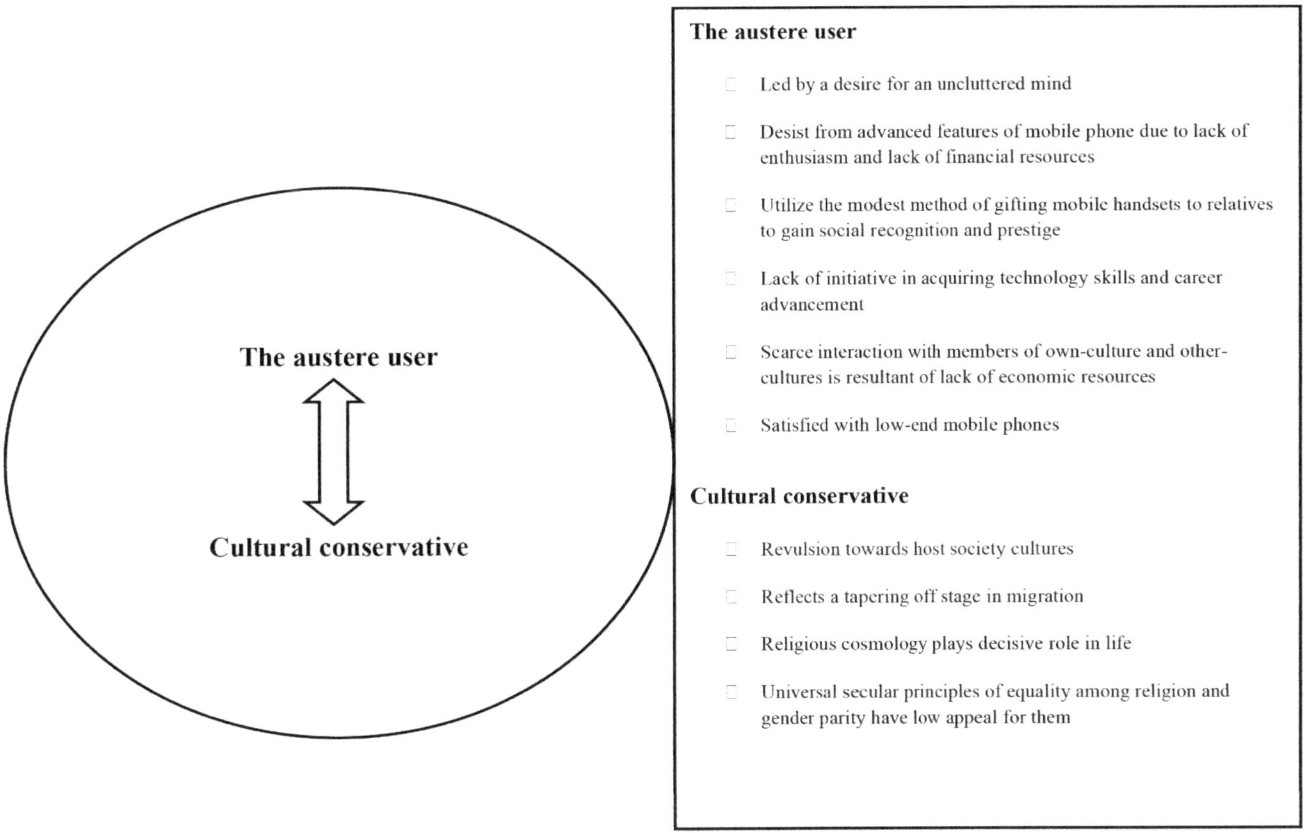

Figure 7. Diagram summarizing the main characteristics of austere user-cultural conservative.
Source: Created by the authors.

APPENDIX C: QUESTIONNAIRE

Please indicate the extent to which you feel "comfortable" or "not comfortable" with each of the following statements, using the 7 point scale:
(1 = "very comfortable"; 2 = "comfortable"; 3 = "ok"; 4 = "neutral"; 5 = "uncomfortable"; 6 = "very uncomfortable"; 7 = "annoyed")

While staying in Singapore, how comfortable are you with the following:

1. Dealing with people in authority.
2. Obeying traffic rules as a pedestrian.
3. Obeying rules in public transport while traveling.
4. Obeying anti-littering law.
5. Obeying anti-drug law.
6. To get along with the overall order and discipline in Singapore.
7. About status given to workers in Singapore.
8. How do you adapt to the family relationships you see in Singapore?
9. Metro culture in Singapore.
10. How comfortable are you with watching festivities of groups other than yours in this city?
11. Your comfort level about the nightlife in Singapore.
12. How comfortable are you with the dress code of women in this city?
13. Your overall comfort level about male-female relationships in this city.
14. Overall world view of this society.
15. Making yourself understood during your stay here.
16. Talking about yourself with others.
17. How comfortable are you with finding your way around in the city?
18. How comfortable are you with communicating with people of a different ethnic group?
19. Communicating in a language other than your mother tongue.
20. Learning a language to know about the culture it represents.

Please indicate the extent to which you "agree" or "disagree" with each of the following statements, using the 5 point scale.
(1 = "totally disagree"; 2 = "somewhat disagree"; 3 = "neutral"; 4 = "somewhat agree"; 5 = "totally agree")

While in Singapore:

21. It is NOT important for me to be fluent either in [ethnic language] or in the languages used in Singapore (Chinese, English, etc).
22. It is important for me to be fluent in both [ethnic language] as well as the languages used in Singapore.
23. It is more important to me to be fluent in [ethnic language] than in languages used in Singapore.
24. Even if I lose my mother tongue eventually, I should be more competent to use languages used in Singapore.
25. I think of myself as [ethnicity].
26. I feel that I am part of [ethnicity].
27. I am proud of being a(n) [ethnicity].
28. I am happy to be a(n) [ethnicity].
29. Being part of [ethnicity] is embarrassing to me.
30. Being a(n) [ethnicity] is uncomfortable for me.
31. Being part of [ethnic] culture makes me feel happy.
32. Being a(n) [ethnicity] makes me feel good.
33. I think I have become a part of Singaporean society.
34. I will be happy to adopt Singaporean culture.
35. Adopting Singaporean culture will be embarrassing to me.
36. I feel that [ethnic] people should maintain their own cultural traditions and not adapt to those of Singapore.
37. I don't want to attend either Singaporean or [ethnic] social activities.
38. I prefer social activities which involve [ethnic] members only.
39. I prefer only those social activities which involve members from other communities in Singapore (Chinese, Malay, etc.) alone.
40. I feel that it is not important for [ethnic] people either to maintain their own cultural traditions or to adopt Singaporean.
41. I feel that [ethnic] people living here should maintain their own cultural traditions but also adopt other cultural traditions in Singapore (Chinese, Malay, etc.).
42. I feel that [ethnic] people should adopt Singaporean cultural traditions and not maintain those of their own.
43. I prefer to have friends only from other communities in Singapore (Chinese, Malay, etc).

44. It is more important to me to be fluent in languages in Singapore than in [ethnic language].
45. I don't want to have either Singaporean or [ethnic] friends.
46. I prefer to have only [ethnic] friends.
47. I prefer social activities which involve people from other communities in Singapore (like Chinese, Malay, etc.) as well as people from my own [ethnic] community.
48. I prefer to have both [ethnic] friends and friends from other local communities in Singapore (like Chinese, Malay, etc.).
49. I have more than 5 close friends from the local communities in Singapore (like Chinese, Malay, etc.).
50. In Singapore, I have more than 5 friends from my own ethnic community.
51. There is room for a variety of cultures in this country.
52. After living in Singapore, I am concerned about losing my cultural identity.

About the organization you work with and about your job:

53. I am willing to put in a great deal of effort beyond that normally expected in order to help my organization be successful.
54. I talk up my organization to my friends as a great organization to work for.
55. I feel very little loyalty to this organization.
56. I would accept almost any type of job assignment in order to keep working for this organization.
57. I find that my values and the organization's values are very similar.
58. I am proud to tell others that I am part of this organization.
59. I could just as well be working for a different organization as long as the type of work was similar.
60. This organization really inspires the very best in me in the way of job performance.
61. It would take very little change in my present circumstances to cause me to leave this organization.
62. I am extremely glad that I chose this organization to work for over others I was considering at the time I joined.
63. There's not too much to be gained by sticking with this organization indefinitely.
64. Often, I find it difficult to agree with this organization's policies on important matters relating to its employees.
65. I really care about the fate of this organization.

66. For me this is the best of all possible organizations for which to work.
67. Deciding to work for this organization was a definite mistake on my part.
68. I feel a sense of pride in working for this organization.
69. I really feel as if the organization's problems are my problems.
70. The most important things that happen to me involve my work.
71. I live, eat, and breathe my job.
72. I have warm feelings toward this organization as a place to work live and work with.
73. I would be quite willing to spend the rest of my career with this organization.
74. I volunteer for tasks that are not required.
75. I complete my assigned duties on time.
76. I comply with the rules and regulations of this organization.
77. I do what my boss says without complaint.
78. Unless I'm rewarded for it in some way, I see no reason to expend extra effort on behalf of this organization.
79. In my job here, I sometimes have to act in ways that are not completely consistent with my true values.
80. I attend functions that are not required, but that help the organization's image.
81. This organization has a tradition of worthwhile accomplishments.
82. Since joining this organization, my personal values and those of the organization have become more similar.
83. At times I wonder how I have begun to feel more committed towards my work after coming to Singapore.

Multiculturalism in workplace:

84. It is a pleasure to work with people of different ethnicities.
85. An organization that has a variety of ethnic and cultural groups is more able to tackle new problems as they occur.
86. It is best for an organization if all workers forget their different ethnic and cultural differences and work in unison.
87. We should recognize that cultural and racial diversity is a fundamental characteristic of my organization.
88. An organization that has a variety of ethnic or cultural groups has more problems with productivity than organizations with one or two basic cultural groups.
89. The productivity of the organization is undermined by workers of different ethnic and cultural backgrounds sticking to their old ways.

For answering the following questions, please follow the key given below: *(1 = at least once a day, 2 = more than once in a week but not everyday, 3 = once in a week, 4 = less than once in a week but at least once in every 15 days, 5 = with a time gap of at least 15 days, 6 = almost never)*

90. How often do you call your family back home?
91. How often do you call your friends back home?
92. How often do you call your friends in other countries?
93. How often do you call your friends from your community in Singapore?
94. How often do you call your friends from other communities in Singapore?
95. How often do you call your fellow-workmen, at or below your level?
96. How often do you call anyone above your status in office?

Please answer the following:
For answering the following questions, please follow the key given below: *(1 = more than one hour, 2 = between 30 mts to 1 hour, 3 = between 10 mts and 29 mts, 4 = approximately 5 to 9 mts, 5 = for a very short time, 6 = not applicable)*

97. On average, for how long do you talk to your family over phone at a time?
98. For how long do you talk to your friends back home at a time over phone?
99. For how long do you talk to your friends in other countries over phone at a time?
100. For how long do you talk to your friends from your own community in Singapore?
101. For how long do you talk to your friends in other communities in Singapore?
102. For how long do you talk to your fellow-workmen, at or below your level?
103. For how long do you talk to someone above your status in office over phone?

End of survey

APPENDIX D: INTERVIEW QUESTIONS

Mobile phone usage history:

1. For how long have you been using the mobile phone?
2. How has your communication over mobile phone to family, friends, and colleagues changed over time, especially during the migration phase?
3. What are the new skills you acquired over time in order to interact with newer and more sophisticated affordances of the phone?

Usage and handling:

1. How has the mobile phone helped you maintain relationship with family and friends?
2. How has the mobile phone helped you manage your life in the host society?
3. How do you use the mobile phone for pastime and distraction?

Prestige and social identity:

1. Have you come across situations when you have been identified by others with a particular pattern of your mobile phone usage?
2. Has your mobile device helped increase your prestige among peers?

Meta-communication:

1. How and what do you communicate with others about your experience of using the mobile phone?
2. How have you come to know about mobile services in the host country?

Acculturation orientation in personal/social lives:

1. Explain the differences in the culture of Singapore from that of your own culture.
2. If there are differences, how do you deal with them?
3. Do you think if you relate to other communities, you will be losing your cultural identity?
4. Do you think you have a good number of friends among the local communities in Singapore? How many? Do you think you've good number of friends from your own ethnic community in Singapore? How many?

5. How does your relationship with the host communities in Singapore affect your identity as a Malayali and/or Indian?
6. How do you maintain a bonding with your family back home with the help of mobile phone?
7. Which all cultural aspects of Singapore are you already familiar with? How similar are they to what you've practiced and learned? *(To verify whether Tamil population feels more inclined to Singapore's situation)*

Communication links with home culture:

1. How often do you talk over mobile phone to your family and friends in India/Bangladesh?
2. How often do you talk over mobile phone to friends and colleagues from other cultures in Singapore?
3. Are you a member of any of the cultural/social service organization in Singapore?
4. Do you attend any of the informal gatherings with co-ethics? How do you maintain communication links with various members of such groups?

Competency in host society language:

1. How competent are you in English?
2. Which language do you use when talking to members of other communities (e.g. Thai, Chinese, etc.), especially over mobile phone?
3. In which language do you speak the most while talking over mobile phone?
4. While speaking in a language other than your mother tongue, is it better to be face to face with your listener or is it better to talk to the listener over mobile phone? Why?

Adaptation to host society rules:

1. Do you think there is a great deal of difference between the rules and regulations in your country of origin and in Singapore? Explain.
2. Do you think you've adapted to the rules and regulations in Singapore?

Overall comfort in host society:

1. Explain how you find it to be working in a culture other than yours.
2. Do you feel any physical or mental discomfort when you adapt to the Singaporean conditions?

3. To what extent do you think you have adapted yourself to the culture of Singapore?

Organizational commitment:

1. Do you think you'll call someone over phone seeking help to do a task better?
2. Do you think that mobile phones can increase your work productivity in the new culture?
3. How much have you adapted to the multicultural workforce of your organization
4. How have mobile phones helped you improve your performance as you cope with your work in Singapore?
5. Did you use mobile phones for any purpose while working in your home country (if yes, in what all ways), (if no, what was the reason for not using a phone)?
6. Explain how you feel when working in a multicultural environment.
7. For what all purposes do you use mobile phone during work?
8. Does your mobile phone help you communicate more with people from other cultures, or does it not? Explain.

APPENDIX E: METHODS IN DETAIL

Factor Analysis 1: Acculturation

The quantitative part of the study had persistent problems with factor analysis. The first task was to identify the patterns in factor loadings. An analysis was done in order to identify the items that loaded on to more than one factor. Only those items that loaded on to more than one factor with a factor loading difference less than 0.15 were included in the analysis.

(i) Item "I don't want to have either Singaporean or [ethnic] friends" had factor loading .49 on factor 2 and .413 on factor 4. This item corresponded to the "separation" construct in acculturation wherein the migrants neither want to relate to other cultures nor want to retain their cultural identity. It did not fit with other questions in factor 2, as the items did not form a meaningful collection of items. Even after removing the item in contention, factor 2 failed to provide a meaningful collection of items. Hence, factor 2 was not included in the final analysis. Grouping the said item on "separation" did not make factor 4 meaningful either. Since the item was altogether removed from analysis, the problem of "close factor loading" did not make a negative impact in analysis.

(ii) Item "I feel that [ethnic group] should adopt the Singaporean cultural traditions and not maintain those of our own" loaded closely on factors 4 (.463) and 5 (.408). Removing it from factor 4 did not make the remaining collection of questions in that factor more meaningful, nor did it improve the Cronbach's alpha of the factor (0.6). When it was included in factor 5, it fit to the factor meaningfully, since all the items in the factor represented the "assimilation" construct, wherein the migrants wanted to relinquish their cultural identity and heritage and adopt the host society culture completely. However, factor 5, even after adding the item in contention, did not have an acceptable internal reliability (.61). Hence, the factor was not included in the statistical model in the subsequent steps of analysis.

(iii) Item "It is more important to me to be fluent in languages in Singapore than in [ethnic language]" loaded both on Factor 4 (.367) and 5 (.44). However, inclusion or exclusion of the item to/from the factors did not help in the analysis: either the factors failed to achieve any meaning considering all grouped items together or the factors lacked reliability. Thus, these factors were omitted from later analysis. The closeness of factor loading of the item was considered a non-issue for the later analysis.

Factor Analysis 2: Adaptation to Social Life

A factor analysis using varimax rotation and unrestricted number of factors was used to determine which items best captured the adaptation outcomes for this sample population. Although five factors emerged following factor analysis, only three factors measuring different aspects of adaptation were used in the study. Remaining two factors did not have a sufficient number of items to form a meaningful scale (refer to table 4 in appendix A for factor loadings, percentage of variance, and internal reliability of factors). All items were reversed before creating the following measures. Combined, the three adaptation outcomes have been termed "adaptation to social life" in the study.

The adaptation to social life scale was created by taking the average of 14 items that grouped in three factors and was internally reliable; Cronbach's alpha = .76.

Other combination of items was also tried in order to identity other meaningful sub-scales. For instance, items "Overall order and discipline in society," "Family relationships in host society," and "Status given to workers in this society" had loaded well on factor 5, even though in the case of the first two items their highest factor loadings corresponded to different factors. Nevertheless, these three items could be meaningfully grouped to form a construct that could represent an additional dimension of social adaptation. However, the internal reliability (Cronbach's alpha = .52) of these three questions was not at acceptable level. In short, although two items—"Overall order and discipline in society" and "Metro culture"—had loaded on to different factors with close factor loadings, they were retained to the factor on which they had the highest factor loading. This helped keep the overall internal reliability of the scale "social adaptation" at the current level, which would have dropped had the items were removed.

Factor Analysis 3: Organizational Commitment

A factor analysis using varimax rotation and unrestricted number of factors was used to determine which items best captured the adaptation outcomes for this sample population. Although six factors emerged, only three factors with eigenvalues above 1.5 were included in the study. Although some items loaded on to more than one factor, none except two such items was removed from the analysis. The removal of the two items (as indicated in table 5 in appendix A) helped improve the internal reliability of the organizational commitment scale. There was no instance of items loading simultaneously on to more than one factor (the closeness in factor loadings has been defined as a "difference of less than 0.15") observed for the factors 4, 5, and 6. However, for factors 1, 2, and 3 a few items loaded on to more than one factor. It implied that constructs compliance, identification, and internalization have not been found clearly differentiated in the sample. However, this overlap of constructs did not influence the later analysis of data, as the reliability of the scale as a whole was found to be good.

Factor Analysis 4: Volume of Mobile Communication

A factor analysis with varimax rotation and unrestricted number of factors was used to determine whether the conceptual foundation distinguishing host and home culture calls was valid in the sample population. Three factors emerged, but they did not reflect the cultural orientation of mobile calling. Hence, the scales were created based on the conceptual foundation of calling to home and host cultures (refer to table 6 in appendix A for factor loadings and reliabilities).

Scales Used in the Survey Instrument

(a) *Cultural identity*: A five-point Likert scale with eight items recorded responses from "1—strongly disagree" to "5—strongly agree." The scale was reliable with a Cronbach's alpha of 0.85.

(b) *Multiculturalism in workplace*: Four items loaded on to the same factor. A five-point Likert scale with four items recorded responses from "1—strongly disagree" to "5—strongly agree." The scale was reliable, Cronbach's alpha: .74.

(c) *Adaptation to social life*: The 5-point Likert scale from the original MIRIPS instrument was enhanced to include a 7-point Likert scale (from 1 "very comfortable" to 7 "annoyed") so as to capture more nuanced insight into the experiences of the interviewees. Similar changes have been made by researchers who adapted the scale to specific cultural contexts. For example, Wilson (2013) re-defined the questionnaire items to make it a "revised sociocultural adaptation scale" (R-SCAS), but retaining the basic dimensions of the scale. Earlier, Ward and Kennedy (1999) prepared and tested a comprehensive SCAS with 41 items (p. 663), but noted that the scale "is a flexible instrument [that] can be easily modified according to the characteristics of the sojourning sample" (p. 662). In a similar manner, modifications were made to the original "socio-cultural competence" scale of MIRIPS as mentioned above. The three underlying constructs in the factor include:

    (i) *Compliance with host society rules:* Five items on rules and regulations loaded on the same factor. Questions included, "While staying in Singapore, how comfortable are you with: obeying rules in public transport?" "Obeying anti-littering law?" and "How comfortable are you with the overall order and discipline in this society?" The last item loaded on to two different factors (Factors 1 and 5), with close factor loadings. However, removing it from Factor 1 reduced the internal reliability of the scale on social adaptation, as well as of the sub-scale "compliance with host society rules." Hence, the item was retained in factor 1.

(ii) *Proficiency in host society language(s):* Six items on language proficiency in host country loaded on the same factor. Questions included, "While staying in Singapore, how comfortable are you with: 'Dealing with people in authority'?" "Communicating with people of a different ethnic group?" and "Communicating in a language other than your mother tongue?"

(iii) *Adaptation to host society's custom and life-style:* Five items on adaptation to host society's custom and life-style loaded on the same factor. Questions included, "Family relationships in the host country," "With metro culture" and "Overall worldview of this society." Item "Family relationships in the country" loaded on three different factors—Factors 3, 4, and 5—but not very closely (i.e., with a difference higher than 0.15). Item "metro culture" loaded on to two factors—Factors 3 and 4—closely. However, removing it from Factor 3 reduced the internal reliability of the scale on social adaptation as a whole, as well as of the sub-scale "adaptation to host society's custom and life-style." Hence, the item was retained in factor 3.

(d) *Organizational commitment*: The 19-item scale used in the study was internally reliable (Cronbach's alpha: .925).

References

Abdullah, N. (2005). Foreign bodies at work: Good, docile and other-ed. *Asian Journal of Social Science, 33*(2), 223–245. doi:10.1163/1568531054930785

Adapa, S. (2002). Birds of 'passage': Migration of south Indian laborers to Southeast Asia. *Critical Asian Studies, 34*(1), 89–115.

Agger, B. (2011). iTime: Labor and life in a smartphone era. *Time & Society, 20*(1), 119–136.

Ahsan Ullah, A. K. M. (2010). *Population migration and Asia: Theories and practice*. Hauppauge, NY: Nova Science.

Alba, R. D., & Nee, V. (2005). *Remaking the American mainstream: Assimilation and contemporary immigration*. Cambridge, MA: Harvard University Press.

Ali, S., & Fokkema, T. (2015). The importance of peers: Assimilation patterns among second-generation Turkish immigrants in Western Europe. *Journal of Ethnic and Migration Studies, 41*(2), 260–283. doi:10.1080/1369183X.2014.921114

Alinejad, D. (2011). Mapping homelands through virtual spaces: Transnational embodiment and Iranian diaspora bloggers. *Global Networks, 11*(1), 43–62. doi:10.1111/j.1471–0374.2010.00306.x

Anderson, B. (1983/2006). *Imagined communities*. London: Verso.

Anderson, C. M., & Martin, M. M. (1995). Why employees speak to coworkers and bosses: Motives, gender, and organizational satisfaction. *Journal of Business Communication, 32*(3), 249–265.

Arasaratnam, L. A. (2008). Acculturation processes of Sri Lankan Tamil immigrants in Sydney: An ethnographic analysis using the Bidirectional Model (BDM). *Australian Journal of Communication, 35*(1), 57–68.

Arends-Tóth, J., & Van De Vijver, F. J. R. (2003). Multiculturalism and acculturation: Views of Dutch and Turkish–Dutch. *European Journal of Social Psychology, 33*(2), 249–266. doi:10.1002/ejsp.143

Aricat, R. G. (2015). Mobile/social media use for political purposes among migrant laborers in Singapore. *Journal of Information Technology & Politics, 12*(1), 18–36. doi:10.1080/19331681.2014.994156

Aricat, R. G., Karnowski, V., & Chib, A. (2015). Mobile phone appropriation and migrant acculturation: A case study of an Indian community in Singapore. *International Journal of Communication, 9*, 2221–2242.

Atabaki, T. (2015). Far from home, but at home: Indian migrant workers in the Iranian oil industry. *Studies in History, 31*(1), 85–114. doi:10.1177/0257643014558477

Bakardjieva, M. (2006). The consumption junction revisited: Networks and contexts. In R. Kraut, M. Brynin, & S. Kiesler (Eds.), *Computers, phones, and the internet: Domesticating information technology*. Oxford Scholarship Online. doi:10.1093/acprof:oso/9780195312805.003.0008

Baker W. E. (1990). Market networks and corporate behavior. *American Journal of Sociology*. 96:589.625

Bangladesh Bureau of Statistics. (2017). *Statistical pocketbook Bangladesh: 2016*. Retrieved from http://bbs.portal.gov.bd/sites/default/files/files/bbs.portal.gov.bd/page/d6556cd1_dc6f_41f5_a766_042b69cb1687/PocketBook2016.pdf

Bar, F., Pisani, F., & Weber, M. (2007). Mobile technology appropriation in a distant mirror: Baroque infiltration, creolization and cannibalism. Retrieved from http://citeseerx.ist.psu.edu/viewdoc/download?doi=10.1.1.184.5809&rep=rep1&type=pdf

Baron, L. F., Neils, M., & Gomez, R. (2014). Crossing new borders: Computers, mobile phones, transportation, and English language among Hispanic day laborers in Seattle, Washington. *Journal of the Association for Information Science and Technology, 65*(1), 98–108. doi:10.1002/asi.22949

Barrett, A., & Mosca, I. (2013). The psychic costs of migration: Evidence from Irish return migrants. *Journal of Population Economics, 26*(2), 483–506.

Barry, D. (2001). Development of a new scale for measuring acculturation: The East Asian Acculturation Measure (EAAM). *Journal of Immigrant & Minority Health, 3*(4), 193–197.

Barry, K. (2006). Home and away: The construction of citizenship in an emigration context. *New York University Law Review, 81*(1), 11–59.

Bauböck, R. (2003). Towards a political theory of migrant transnationalism. *International Migration Review, 37*(3), 700–723.

Benitez, J. L. (2006). Transnational dimensions of the digital divide among Salvadoran immigrants in the Washington DC metropolitan area. *Global Networks, 6*(2), 181–199. doi:10.1111/j.1471–0374.2006.00140.x

Bernal, V. (2006). Diaspora, cyberspace, and political imagination: The Eritrean diaspora online. *Global Networks, 6*(2), 161–179.

Berry, J. W. (1980). Acculturation as varieties of adaptation. In A. M. Padilla (Ed.), *Acculturation: Theory, models and some new findings* (pp. 9–25). Washington, DC: Westview.

Berry, J. W. (1997). Immigration, acculturation, and adaptation. *Applied Psychology: An International Review, 46*(1), 5–34.

Berry, J. W. (2001). A psychology of immigration. *Journal of Social Issues, 57*(3), 615–631.

Berry, J. W. (2005). Acculturation: Living successfully in two cultures. *International Journal of Intercultural Relations, 29*(6), 697–712.

Berry, J. W. (2010). Mutual Intercultural Relations in Plural Societies (MIRIPS). Retrieved from http://cacr.victoria.ac.nz/projects/research-projects/mirips

Berry, J. W., Kim, U., Power, S., Yong, M., & Bujaki, M. (1989). Acculturation attitudes in plural societies. *Applied Psychology: An International Review, 38*, 185–206.

Bhatia, S., & Ram, A. (2009). Theorizing identity in transnational and diaspora cultures: A critical approach to acculturation. *International Journal of Intercultural Relations, 33*(2), 140–149. doi:10.1016/j.ijintrel.2008.12.009

Bhugra, D., & Jones, P. (2001). Migration and mental illness. *Advances in Psychiatric Treatment, 7*(3), 216–223.

Bijker, W. E., Hughes, T. P., & Pinch, T. J. (2012). *The social construction of technological systems: New directions in the sociology and history of technology*. Cambridge, MA: MIT Press.

Birman, D., Trickett, E., & Buchanan, R. M. (2005). A tale of two cities: Replication of a study on the acculturation and adaptation of immigrant adolescents from the former Soviet Union in a different community context. *American Journal of Community Psychology, 35*(1–2), 83–101.

Bisin, A., Patacchini, E., Verdier, T., & Zenou, Y. (2008). Are Muslim immigrants different in terms of cultural integration? *Journal of the European Economic Association, 6*(2–3), 445–456.

Boase, J., & Ling, R. (2013). Measuring mobile phone use: Self-report versus log data. *Journal of Computer-Mediated Communication, 18*(4), 508–519. doi:10.1111/jcc4.12021

Bourhis, R. Y., Moise, L. C., Perreault, S., & Senecal, S. (1997). Towards an Interactive Acculturation Model: A social psychological approach. *International Journal of Psychology, 32*(6), 369–386.

Boyatzia, R. E. (1998). *Transforming qualitative information: Thematic analysis and code*. London, UK: Sage.

Breeding, M. E. (2016). The micropolitics of Indian recruitment agencies: India-Gulf migration from a local Indian perspective. In S. Irudaya Rajan (Ed.), *India migration report 2014: Diaspora and development* (pp. 184–201). London: Routledge.

Broeders, D., & Hampshire, J. (2013). Dreaming of seamless borders: ICTs and the pre-emptive governance of mobility in Europe. *Journal of Ethnic & Migration Studies, 39*(8), 1201–1218. doi:10.1080/1369183X.2013.787512

Browne, S. (2010). Digital epidermalization: Race, identity and biometrics. *Critical Sociology, 36*(1), 131–150. doi:10.1177/0896920509347144

BTRC. (2014). *Mobile phone subscribers in Bangladesh*. Bangladesh Telecommunication Regulatory Commission. Retrieved from http://www.btrc.gov.bd/

Buchanan II, B. (1974). Building organizational commitment: The socialization of managers in work organizations. *Administrative Science Quarterly, 19*(4), 533–546.

Burrell, J., & Anderson, K. (2008). "I have great desires to look beyond my world": Trajectories of information and communication technology use among Ghanaians living abroad. *New Media & Society, 10*(2), 203–224. doi:10.1177/1461444807086472

Cai, D. A., & Rodríguez, J. I. (1997). Adjusting to cultural differences: The intercultural adaptation model. *Intercultural Communication Studies*, *6*, 31–42.

Cartier, C., Castells, M., & Qiu, J. L. (2005). The information have-less: Inequality, mobility, and translocal networks in Chinese cities. *Studies in Comparative International Development, 40*(2), 9–34.

Chae, M. H., & Foley, P. F. (2010). Relationship of ethnic identity, acculturation, and psychological well-being among Chinese, Japanese, and Korean Americans. *Journal of Counseling & Development, 88*(4), 466–476.

Chan, A. (2011, Feb). Hired on sufferance: China's migrant workers in Singapore. *China Labour Bulletin*. Retrieved from http://www.clb.org.hk/content/hired-sufferance-china%E2%80%99s-migrant-workers-singapore

Chan, S. H., & Qiu, H. H. (2011). Loneliness, job satisfaction, and organizational commitment of migrant workers: Empirical evidence from China. *International Journal of Human Resource Management, 22*(5), 1109–1127. doi:10.1080/09585192.2011.556785

Chang, R. (2010, July 5). Life is tough at 24 with us and them. *The Straits Times.*

Chatman, J. A., Polzer, J. T., Barsade, S. G., & Neale, M. A. (1998). Being different yet feeling similar: The influence of demographic composition and organizational culture on work processes and outcomes. *Administrative Science Quarterly, 43*(4), 749–780.

Chen, V. H-H. (2014). Facilitating social inclusion of migrant workers through digital game play. *Proceedings of the Annual Workshop of the AIS Special Interest Group for ICT in Global Development*. Retrieved from http://aisel.aisnet.org/cgi/viewcontent.cgi?article=1004&context=globdev2014

Chen, W. (2010). Internet-usage patterns of immigrants in the process of intercultural adaptation. *CyberPsychology, Behavior & Social Networking, 13*(4), 387–399. doi:10.1089/cyber.2009.0249

Cheng, C.-Y., & Lee, F. (2009). Multiracial identity integration: Perceptions of conflict and distance among multiracial individuals. *Journal of Social Issues, 65*(1), 51–68. doi:10.1111/j.1540-4560.2008.01587.x

Cheong, K. (2015). *Kwento: Designing a mobile application to support family storytelling for domestic helpers in Singapore.* Unpublished Masters Thesis, National University of Singapore, Singapore.

Chia, Z. I. (2014). Filipino Independence Day plans reveal political rifts in Singapore. *Asian Correspondent.* Retrieved from https://asiancorrespondent.com/2014/04/filipino-independence-day-celebrations-reveal-political-rifts-in-singapore/#3x9j15AVOL4xIRfO.97

Chib, A., & Aricat, R. (2012). Seeking the non-developmental within the developmental: Mobile phones in the globalized migration context. In R. Parker & R. P. Appelbaum (Eds.), *Can emerging technologies make a difference in development?* (pp. 153–167). New York: Routledge.

Chib, A., Malik, S., Aricat, R. G., & Kadir, S. Z. (2014). Migrant mothering and mobile phones: Negotiations of transnational identity. *Mobile Media & Communication*, *2*(1), 73–93. doi:10.1177/2050157913506007

Chib, A., Wilkin, H. A., & Mei Hua, S. R. (2013). International migrant workers' use of mobile phones to seek social support in Singapore. *Information Technologies & International Development, 9*(4), 19–34.

Chin, Y. (2008). *Foreign workers in Singapore: Integrating or segregating them.* Rajaratnam School of International Studies. Retrieved from http://www.rsis.edu.sg/rsis-publication/cens/1129-foreign-workers-in-singaporei/#.VE9NViKUfD4

Chiswick, B. R., & Miller, P. W. (2007). Computer usage, destination language proficiency and the earnings of natives and immigrants. *Review of Economics of the Household, 5*(2), 129–157. doi:10.1007/s11150–007–9007–0

Chouliaraki, L., & Fairclough, N. (1999). *Discourse in late modernity: Rethinking critical discourse analysis*. Edinburgh: Edinburgh University Press.

Christiansen, C. C. (2004). News media consumption among immigrants in Europe: The relevance of diaspora. *Ethnicities, 4*(2), 185–207. doi:10.1177/1468796804042603

Chua, B. H. (2003). Multiculturalism in Singapore: An instrument of social control. *Race & Class, 44*(3), 58–77. doi:10.1177/0306396803044003025

CIA World Factbook. (2017). *India*. Retrieved from https://www.cia.gov/library/publications/the-world-factbook/geos/in.html

Citron, J. L. (1995). Can cross-cultural understanding aid second language acquisition? Toward a theory of ethno-lingual relativity. *Hispania, 78*(1), 105–113.

Ciunova-Shuleska, A., Grishin, M., & Palamidovska, N. (2012). Young people's attitudes toward mobile phones in the Republic of Macedonia. *Economic Research, 25*(4), 1067–1078.

Claus, L., Lungu, A. P., & Bhattacharjee, S. (2011). The effects of individual, organizational and societal variables on the job performance of expatriate managers. *International Journal of Management, 28*(1), 249–271.

Coe, N. M., & Kelly, P. F. (2000). Distance and discourse in the local labour market: The case of Singapore. *Area, 32*(4), 413–422.

Coffey, D., Papp, J., & Spears, D. (2015). Short-term labor migration from rural north India: Evidence from new survey data. *Population Research and Policy Review, 34*(3), 361–380.

Cohen, E. H. (2011). Impact of the group of co-migrants on strategies of acculturation: Towards an expansion of the Berry model. *International Migration, 49*(4), 1–22. doi:10.1111/j.1468–2435.2009.00589.x

Cowan, R. S. (1976). The Industrial Revolution in the home. *Technology and Culture, 17*(1), 1–23.

Cox, T., & Finley-Nickelson, J. (1991). Models of acculturation for intra-organizational cultural diversity. *Canadian Journal of Administrative Sciences, 8*(2), 90–101.

Creswell, J. W. (2007). *Qualitative inquiry & research design: Choosing among five approaches*. Thousand Oaks: Sage.

Croucher, S. M., & Kramer, E. (2017). Cultural fusion theory: An alternative to acculturation. *Journal of International and Intercultural Communication, 10*(2), 97–114. doi:10.1080/17513057.2016.1229498

Croucher, S. M., Oommen, D., & Steele, E. L. (2009). An examination of media usage among French-Muslims. *Journal of Intercultural Communication Research, 38*(1), 41–57. doi:10.1080/17475750903478113

Crul, Maurice, and Jens Schneider. (2009). The second generation in Europe: Education and the transition to the labour market. Washington, DC: Migration Policy Institute.

Dahlberg, L. (2007). Rethinking the fragmentation of the cyberpublic: From consensus to contestation. *New Media & Society, 9*(5), 827–847. doi:10.1177/1461444807081228

Dandy, J., & Pe-Pua, R. (2010). Attitudes to multiculturalism, immigration and cultural diversity: Comparison of dominant and non-dominant groups in three Australian states. *International Journal of Intercultural Relations, 34*(1), 34–46.

Davis, C. P. (2004). Beyond Miami: The ethnic enclave and personal income in various Cuban communities in the United States. *International Migration Review, 38*(2), 450–469. doi:10.1111/j.1747–7379.2004.tb00205.x

De Fleur, M. L., & Cho, C. S. (1957). Assimilation of Japanese born women in an American city. *Social Problems, 4*, 244–257.

de Souza e Silva, A., Sutko, D. M., Salis, F. A., & de Souza e Silva, C. (2011). Mobile phone appropriation in the favelas of Rio de Janeiro, Brazil. *New Media & Society, 13*(3), 411–426. doi:10.1177/1461444810393901

Dekker, R., & Engbersen, G. (2014). How social media transform migrant networks and facilitate migration. *Global Networks, 14*(4), 401–418. doi:10.1111/glob.12040

Delgado-Wise, R. (2014). A critical overview of migration and development: The Latin American challenge. *Annual Review of Sociology, 40*, 643–663. doi:10.1146/annurev-soc-071811–145459

Denzin, N. K. (1992). *Symbolic interactionism and cultural studies*: Oxford, UK: Blackwell, 1992.

Department of Statistics Malaysia. (2017). Malaysia survey report 2016. Retrieved from https://www.dosm.gov.my/v1/index.php?r=column/pdfPrev&id=RldLdXFnQUppeGM5Yk1HODlTMVl6UT09

Diminescu, D. (2008). The connected migrant: An epistemological manifesto. *Social Science Information, 47*(4), 565–579. doi:10.1177/0539018408096447

Donner, J. (2006). The social and economic implications of mobile telephony in Rwanda: An ownership/access typology. *Knowledge Technology & Policy, 19*(2), 17–28.

Donner, J. (2007). The rules of beeping: Exchanging messages via intentional "missed calls" on mobile phones. *Journal of Computer-Mediated Communication, 13*(1), 1–22. doi:10.1111/j.1083–6101.2007.00383.x

Donner, J. (2008). Research approaches to mobile use in the developing world: A review of the literature. *The Information Society, 24*(3), 140–159.

Donner, J. (2009). Blurring livelihoods and lives: The social uses of mobile phones and socioeconomic development. *Innovations: Technology, Governance, Globalization, 4*(1), 91–101.

Dubois, E., & Gaffney, D. (2014). The multiple facets of influence: Identifying political influentials and opinion leaders on Twitter. *American Behavioral Scientist, 58*(10), 1260–1277. doi:10.1177/0002764214527088

Durkheim, E. (1897/1951). *Suicide: A study in sociology*. New York: The Free Press.

Ellul, J. (1964). *The technological society*. New York: Vintage Books.

Ely, R. J., & Thomas, D. A. (2001). Cultural diversity at work: The effects of diversity perspectives on work group processes and outcomes. *Administrative Science Quarterly, 46*(2), 229–273.

Emmer, P. C. (Ed.). (1986). *Colonialism and migration; indentured labour before and after slavery*. Dordrecht: Martinus Nijhoff; Springer Netherlands.

Eriksen, T. H. (1996). Ethnicity, race, class and nation. In J. Hutchinson & A.D. Smith (Eds.), *Ethnicity* (pp. 28–34). Oxford: Oxford University Press.

Essam, A. H. M. (2015). Information needs of local domestic workers in the Arab Republic of Egypt. *The Electronic Library, 33*(4), 643–667. doi:10.1108/EL-01–2014–0012

Eveland, W. P., Hayes, A. F., Shah, D. V., & Kwak, N. (2005). Understanding the relationship between communication and political knowledge: A model comparison approach using panel data. *Political Communication, 22*(4), 423–446.

Fairclough, N. (2003). *Analysing discourse: Textual analysis for social research*. New York: Routledge.

Fathi, A. (1973). Mass media and a Moslem immigrant community in Canada. *Anthropologica, 15*(2), 201–230.

Fink, C., & Kenny, C. J. (2003). *W(h)ither the digital divide*. Washington, DC: World Bank.

Fitzsimmons, S. R. (2013). Multicultural employees: A framework for understanding how they contribute to organizations. *Academy of Management Review, 38*(4), 525–549. doi:10.5465/amr.2011.0234

Ford, R. (2011). Acceptable and unacceptable immigrants: How opposition to immigration in Britain is affected by migrants' region of origin. *Journal of Ethnic & Migration Studies, 37*(7), 1017–1037. doi:10.1080/1369183x.2011.572423

Fortunati, L., Manganelli, A., Law, P.-L., & Yang, S. (2008). Beijing calling . . . mobile communication in contemporary China. *Knowledge, Technology & Policy, 21*(1), 19–27. doi:10.1007/s12130–008–9040–1

Foucault, M. (1980). *Power/knowledge: Selected interviews and other writings 1972–1977* (C. Gordon, L. Marshall, J. Mepham, & K. Soper, Trans.). New York: Pantheon Books.

Foucault, M. (1981). The order of discourse. In R. Young (Ed.), *Untying the text: A post-structuralist reader* (pp. 51–78). Boston: Routledge & Kegan Paul.

Foucault, M. (1991). Politics and the study of discourse. In G. Burchell, C. Gordon, & P. Miller (Eds.), *The Foucault effect: Studies in governmentality* (pp. 53–72). Chicago: University of Chicago Press.

Foucault, M. (1997). What is enlightenment? In P. Rabinow (Ed.), *The essential works of Foucault, 1954–1984 (Vol. 1): Ethics: subjectivity and truth* (pp. 303–319). New York: New Press.

Fowler, F. J. J. (1992). How unclear terms affect survey data. *The Public Opinion Quarterly, 56*(2), 218–231. doi:10.2307/2749171

Frankenberg, E., Kupper, K., Wagner, R., & Bongard, S. (2013). Immigrant youth in Germany: Psychological and sociocultural adaptation. *European Psychologist, 18*(3), 158–168. doi:10.1027/1016–9040/a000154

Gamburd, M. (2009). Advocating for Sri Lankan migrant workers. *Critical Asian Studies, 41*(1), 61–88.

García-Montes, J. M., Caballero-Muñoz, D., & Pérez-Álvarez, M. (2006). Changes in the self resulting from the use of mobile phones. *Media, Culture & Society, 28*(1), 67–82. doi:10.1177/0163443706059287

Garrett, K. R., Bimber, B., Gil de Zuniga, H., Heinderyckx, F., Kelly, J., & Smith, M. (2012). Transnational connections: New ICTs and the study of political communication. *International Journal of Communication, 6.*

Geser, H. (2006). Is the cell phone undermining the social order?: Understanding mobile technology from a sociological perspective. *Knowledge, Technology & Policy, 19*(1), 8–18.

Gillan, M. (2002). Refugees or infiltrators? The Bharatiya Janata Party and "illegal" migration from Bangladesh. *Asian Studies Review, 26*(1), 73–95.

Glover, C., & Sim, L. (2002). *Entrepreneur immigrant survey: A mail survey of entrepreneur immigrants in Vancouver, Canada, post-terms and conditions removal.* Ottawa: Strategic Research and Review, CIC, and International Consortium for Venture Expertise, University of Victoria.

Goh, C. L., Yong, C., & Sim, W. (2013, January 31). White paper on population: Reactions—Coping with a rise in numbers; S'poreans' fears, *The Straits Times.*

Goh, D. P. S. (2013). Multicultural carnivals and the politics of the spectacle in global Singapore. *Inter-Asia Cultural Studies, 14*(2), 228–251.

Goldlust, J., & Richmond, A. (1974). A multivariate model of immigrant adaptation. *International Migration Review, 8*, 193–225.

Gordon, M. M. (1964). *Assimilation in American life: The role of race, religion and national origins.* New York: Oxford University Press.

Graves, T. (1967). Acculturation, access, and alcohol in a tri-ethnic community. *American Anthropologist, 69*(3–4), 306–321.

Green, G., & Lockley, E. G. (2012). Communication practices of the Karen in Sheffield: Seeking to navigate their three zones of displacement. *Asian Journal of Communication, 22*(6), 566–583. doi:10.1080/01292986.2012.717095

Griffith, D. M., Childs, E. L., Eng, E., & Jeffries, V. (2007). Racism in organizations: The case of a county public health department. *Journal of Community Psychology, 35*(3), 287–302.

Grint, K., & Woolgar, S. (1997). *The machine at work.* Cambridge: Polity.

Grzymala-KazLowska, A. (2005). From ethnic cooperation to in-group competition: Undocumented Polish workers in Brussels. *Journal of Ethnic and Migration Studies, 31*(4), 675–697.

GSMA Intelligence. (2016). Asia Pacific to add 600 million new mobile subscribers by 2020, finds latest GSMA mobile economy study. Retrieved from http://www.gsma.com/newsroom/press-release/asia-pacific-add-600-million-new-mobile-subscribers-2020-finds-latest-gsma-mobile-economy-study/

GSMA Intelligence. (2017). *The mobile economy 2017.* Retrieved from https://www.gsmaintelligence.com/research/?file=9e927fd6896724e7b26f33f61db5b9d5&download

Guarnizo, L. E., Portes, A., & Haller, W. (2003). Assimilation and transnationalism: Determinants of transnational political action among contemporary migrants. *American Journal of Sociology, 108*(6), 1211–1248.

Guo, C., & Al Ariss, A. (2015). Human resource management of international migrants: Current theories and future research. *The International Journal of Human Resource Management, 26*(10), 1287–1297. doi:10.1080/09585192.2015.1011844

Hall, S. (1980). Encoding/decoding. In S. Hall et al. (Eds.), *Culture, media, language* (pp. 128–138). London: Hutchinson.

Hall, S. (1991). Old and new identities, old and new ethnicities. In A. D. King (Ed.), *Culture, globalization, and the world-system: Contemporary conditions for the representation of identity* (pp. 41–68). Binghamton, NY: State University of New York.

Hall, S. (1992). The question of cultural identity. In S. Hall, D. Held, & T. McGrew (Eds.), *Modernity and its futures* (pp. 274–325). Cambridge: Polity Press.

Hall, S. (1996). The new ethnicities. In J. Hutchinson & A. D. Smith (Eds.), *Ethnicity* (pp. 161–164). Oxford: Oxford University Press.

Hamel, J.-Y. (2009). Information and communication technologies and migration. *Human Development Report*: *United Nations*. Retrieved from http://hdr.undp.org/sites/default/files/hdrp_2009_39.pdf

Hamid, W. (2015). Feelings of home amongst Tamil migrant workers in Singapore's Little India. *Pacific Affairs, 88*(1), 5–25. doi:10.5509/20158815

Hammersley, M. (1997). On the foundations of critical discourse analysis. *Language & Communication, 17*(3), 237–248. doi:10.1016/S0271–5309(97)00013-X

"Hanoi plan to ban motorbikes by 2030 to combat pollution." (2017). BBC. Retrieved from http://www.bbc.com/news/world-asia-40498052

Hao, X., & Zhu, R. (2005). Media use by immigrants in the process of cultural adaptation. *Asian Communication Research, 1*.

Haverila, M. (2011). Behavioral aspects of cell phone usage among youth: An exploratory study. *Young Consumers, 12*(4), 310–325. doi:10.1108/17473611111185869

Hernandez, M. Y. (2009). Psychological theories of immigration. *Journal of Human Behavior in the Social Environment, 19*(6), 713–729. doi:10.1080/10911350902910898

Hewison, K., & Young, K. (2006). Introduction: Globalization and migrant workers in Asia. In K. Hewison & K. Young (Eds.), *Transnational migration and work in Asia* (pp. 1–11). Abingdon, Oxon: Routledge.

Hicks, L. E. (1970). Some properties of ipsative, normative, and forced-choice normative measures. *Psychological Bulletin, 74*(3), 167–184. doi:10.1037/h0029780

Holmes, P., & Janson, A. (2008). Migrants' communication practices with ICTs: Tools for facilitating migration and adaptation? *International Journal of Technology, Knowledge and Society, 4*(6), 51–62.

Hood, J. N., & Koberg, C. S. (1994). Patterns of differential assimilation and acculturation for women in business organizations. *Human Relations, 47*(2), 159–181. doi:10.1177/001872679404700202

Horenczyk, G., & Munayer, S. J. (2007). Acculturation orientations toward two majority groups: The case of Palestinian Arab Christian adolescents in Israel. *Journal of Cross-Cultural Psychology*. *38*(76), 77–87.

Horst, H. A. (2006). The blessings and burdens of communication: Cell phones in Jamaican transnational social fields. *Global Networks, 6*(2), 143–159. doi:10.1111/j.1471–0374.2006.00138.x

Horst, H., & Miller, D. (2005). From kinship to link-up: Cell phones and social networking in Jamaica. *Current Anthropology, 46*(5), 755–778. doi:10.1086/432650

Howarth, C., Wagner, W., Magnusson, N., & Sammut, G. (2014). "It's only other people who make me feel Black": Acculturation, identity, and agency in a multicultural community. *Political Psychology, 35*(1), 81–95. doi:10.1111/pops.12020

Hua, S. R. M. (2014). New Singapore law will not keep peace in Little India. *The Diplomat*. Retrieved from http://thediplomat.com/2014/02/new-singapore-law-will-not-keep-peace-in-little-india/

Hutchby, I. (2001). Technologies, texts and affordances. *Sociology, 35*(2), 441–456. doi:10.1177/s0038038501000219

Hutchinson, J. (1994). Cultural nationalism and moral regeneration. In J. Hutchinson & A. D. Smith (Eds.), *Nationalism* (pp. 122–131). New York: Oxford University Press.

Hwang, B.-h., & He, Z. (1999). Media uses and acculturation among Chinese immigrants in the USA. *International Communication Gazette, 61*(1), 5–22. doi:10.1177/0016549299061001001

IMDA. (2017). Statistics on Telecom Services for 2016 (Jan—Jun). *Infocomm Media Development Authority of Singapore*. Retrieved from https://www.imda.gov.sg/

International Organization for Migration. (2018). *World migration report 2018*. Retrieved from http://publications.iom.int/about-iom-publications

International Telecommunication Union. (2015). *ICT facts and figures: The world in 2015*. Retrieved from https://www.itu.int/en/ITU-D/Statistics/Documents/facts/ICTFactsFigures2015.pdf

Jablin, E. M., & Sussman, L. (1983). Organizational group communication: A review of the literature and model of the process. In H. H. Greenbaum, R. L. Falcione, & S. A. Hellweg (Eds.), *Organizational communication: Abstracts, analysis & overview* (pp. 11–50). Beverly Hills, CA: Sage.

Jacobs, A. (2012, July 27). In Singapore, vitriol against Chinese newcomers. *The New York Times*. Retrieved from http://mobile.nytimes.com/2012/07/27/world/asia/in-singapore-vitriol-against-newcomers-from-mainland-china.html

Jamison, A., & Hard, M. (2003). The story-lines of technological change: Innovation, construction and appropriation. *Technology Analysis & Strategic Management, 15*(1), 81–91. doi:10.1080/0953732032000046060

Jasinskaja-Lahti, I., & Liebkind, K. (1999). Exploration of the ethnic identity among Russian-speaking immigrant adolescents in Finland. *Journal of Cross-Cultural Psychology, 30*(4), 527–539. doi:10.1177/0022022199030004007

Jeffres, L. W. (2000). Ethnicity and ethnic media use: A panel study. *Communication Research, 27*(4), 496–535. doi:10.1177/009365000027004004

Jeffres, L. W., & Hur, K. K. (1981). Communication channels within ethnic groups. *International Journal of Intercultural Relations, 5*(2), 115–132.

Jones, G. W. (2012). Population policy in a prosperous city-state: Dilemmas for Singapore. *Population and Development Review, 38*(2), 311–336.

Jurcik, T., Ahmed, R., Yakobov, E., Solopieieva-Jurcikova, I., & Ryder, A. G. (2013). Understanding the role of the ethnic density effect: Issues of acculturation, discrimination and social support. *Journal of Community Psychology, 41*(6), 662–678. doi:10.1002/jcop.21563

Kalba, K. (2008). The adoption of mobile phones in emerging markets: Global diffusion and the rural challenge. *International Journal of Communication 2*, 631–661.

Kanchana, R. (2016). Are India's policies increasing the vulnerability of its female migrants in the Arab Gulf countries? In S. Irudaya Rajan (Ed.), *India migration report 2014: Diaspora and development* (pp. 90–105). London: Routledge.

Karnowski, V., von Pape, T., & Wirth, W. (2008). After the digital divide? An appropriation perspective on the generational mobile-phone divide. In M. Hartmann, P. Rössler, & J. Höflich (Eds.), *After the mobile phone? Social changes and the development of mobile communication* (pp. 185–202). Berlin: Frank & Timme.

Kathiravelu, L. (2012). Social networks in Dubai: Informal solidarities in an uncaring state. *Journal of Intercultural Studies, 33*(1), 103–119. doi:10.1080/07256868.2012.633319

Kathiravelu, L. (2016). *Migrant Dubai: Low wage workers and the construction of a global city*. New York: Palgrave Macmillan.

Katz, J. E. (2006). Mobile communication and the transformation of daily life: The next phase of research on mobiles. *Knowledge, Technology & Policy, 19*(1), 63–71.

Katz, J. E., & Aakhus, M. (Eds.) (2002). *Perpetual contact. Mobile communication, private talk, public performance*. Cambridge, MA: Cambridge University Press.

Kaur, A. (2010). Labour migration in Southeast Asia: Migration policies, labour exploitation and regulation. *Journal of the Asia Pacific Economy, 15*(1), 6–19.

Keck, M. E., & Sikkink, K. (1998). *Activists beyond borders: Advocacy networks in international politics*. Ithaca, NY: Cornell University Press.

Kelman, H. C. (1958). Compliance, identification, and internalization: Three processes of attitude change. *Journal of Conflict Resolution, 2*(1), 51–60.

Kim, J. K. (1980). Explaining acculturation in a communication framework: An empirical test. *Communication Monographs, 47*(3), 155–179.

Kim, M.-J., & Park, J. (2013). Mobile phone purchase and usage behaviours of early adopter groups in Korea. *Behaviour & Information Technology, 33*(7), 693–703. doi:10.1080/0144929X.2013.796001

Kim, S. H. (2012). Max Weber. In E. N. Zalta (Ed.), *The Stanford encyclopedia of philosophy* (Fall 2012 ed.). Retrieved from http://plato.stanford.edu/cgi-bin/encyclopedia/archinfo.cgi?entry=weber

Kim, Y. Y. (1977). Communication patterns of foreign immigrants in the process of acculturation. *Human Communication Research, 4*, 66–77.

Kim, Y. Y. (1978). A communication approach to the acculturation process: A study of Korean immigrants in Chicago. *International Journal of Intercultural Relations, 2*(2), 197–224.

Kim, Y. Y. (1984). Searching for creative integration. In W. B. Gudykunst & Y. Y. Kim (Eds.), *Methods for intercultural communication research* (pp. 13–30). Beverly Hills, CA: Sage.

Kim, Y. Y. (1989). Intercultural adaptation. In M. K. Asante, W. B. Gudykunst, & E. Newmark (Eds.), *Handbook of international and intercultural communication* (pp. 275–294). Newbury Park, CA: Sage Publications.

Kim, Y. Y. (2001). *Becoming intercultural: An integrative theory of communication and cross-cultural adaptation*. Thousand Oaks, CA: Sage.

Kim, Y. Y., & McKay-Semmler, K. (2013). Social engagement and cross-cultural adaptation: An examination of direct- and mediated interpersonal communication activities of educated non-natives in the United States. *International Journal of Intercultural Relations, 37*(1), 99–112. doi:10.1016/j.ijintrel.2012.04.015

Kramer, E. M. (2000). *Ressentiment* and racism: A Nietzschean reading of African American and Korean American conflict. In M. K. Asante & E. Min (Eds.), *Socio-cultural Conflict between African and Korean Americans* (pp. 34–62). New York: University Press of America.

Krishnan, A., & Berry, J. W. (1992). Acculturative stress and acculturation attitudes among Indian immigrants to the United States. *Psychology and Developing Societies, 4*(2), 187–212.

Kumar, A. (2015). Depoliticising illegal immigration from Bangladesh to India. *Strategic Analysis, 39*(1), 22–26. doi:10.1080/09700161.2014.980543

LaFromboise, T., Coleman, H. L., & Gerton, J. (1993). Psychological impact of biculturalism: Evidence and theory. *Psychological Bulletin, 114*(3), 395–412. doi:10.1037/0033–2909.114.3.395

Laguerre, M. S. (2005). Homeland political crisis, the virtual diasporic public sphere, and diasporic politics. *Journal of Latin American Anthropology, 10*(1), 206–225. doi:10.1525/jlca.2005.10.1.206

Landolt, P. (2001). Salvadoran economic transnationalism: Embedded strategies for household maintenance, immigrant incorporation, and entrepreneurial expansion. *Global Networks, 1*(3), 217–241.

Larkey, L. K. (1996). Toward a theory of communicative interactions in culturally diverse workgroups. *Academy of Management Review, 21*(2), 463–491.

Law, P., & Peng, Y. (2007). Cellphones and the social lives of migrant workers in southern China. In R. Pertierra (Ed.), *The social construction and usage of communication technologies: Asian and European experiences* (pp. 126–142). Quezon City: University of the Philippines Press.

Law, P.-l., & Chu, W.-c. R. (2008). ICTs and migrant workers in contemporary China. *Knowledge, Technology & Policy, 21*(2), 43–45. doi:10.1007/s12130–008–9049–5

Law, P.-l., & Peng, Y. (2008). Mobile networks: Migrant workers in southern China. In J. E. Katz (Ed.), *Handbook of mobile communication studies* (pp. 55–64). Cambridge, MA: MIT Press.

Lee, T. (2005). Internet control and auto-regulation in Singapore. *Surveillance & Society, 3*(1), 74–95. Retrieved from http://www.surveillance-and-society.org/Articles3(1)/singapore.pdf

Lefevre, A. S. (2017). Thailand's new labor rules send thousands of migrant workers fleeing. *Reuters*. Retrieved from https://www.reuters.com/article/us-thailand-migrants/thailands-new-labor-rules-send-thousands-of-migrant-workers-fleeing-idUSKBN19O0B6

Leifsen, E., & Tymczuk, A. (2012). Care at a distance: Ukrainian and Ecuadorian transnational parenthood from Spain. *Journal of Ethnic and Migration Studies, 38*(2), 219–236. doi:10.1080/1369183X.2012.646419

Lenard, P. T., & Straehle, C. (2010). Temporary labour migration: Exploitation, tool of development, or both? *Policy and Society, 29*(4), 283–294. doi:10.1016/j.polsoc.2010.09.006

Leonardi, P. M. (2003). Problematizing "new media": Culturally based perceptions of cell phones, computers, and the internet among United States Latinos. *Critical Studies in Media Communication, 20*(2), 160–179.

Leong, C.-H., & Berry, J. W. (2010). *Intercultural relations in Asia: Migration and work effectiveness*. Singapore: World Scientific.

Lepp, A., Li, J., Barkley, J. E., & Salehi-Esfahani, S. (2015). Exploring the relationships between college students' cell phone use, personality and leisure. *Computers in Human Behavior, 43*, 210–219. doi:10.1016/j.chb.2014.11.006

Lévi-Strauss, C. (1966). *The savage mind*. Chicago: University of Chicago Press.

Levitt, P., & Glick Schiller, N. (2004). Conceptualizing simultaneity: A transnational social field perspective on society. *International Migration Review, 38*(3), 1002–1039.

Levitt, P., DeWind, J., & Vertovec, S. (2003). International perspectives on transnational migration: An introduction. *International Migration Review, 37*(3), 565–575.

Ley, D. (2006). Explaining variations in business performance among immigrant entrepreneurs in Canada. *Journal of Ethnic and Migration Studies, 32*(5), 743–764. doi:10.1080/13691830600704123

Licoppe, C. (2004). "Connected" presence: The emergence of a new repertoire for managing social relationships in a changing communication technoscape. *Environment and Planning D: Society and Space, 22*(9), 135–156.

Lievrouw, L., & Livingstone, S. (Eds.). (2002). *Handbook of new media: Social shaping and social consequences*. London, UK: Sage.

Lin, C. A. (2003). An interactive Communication Technology Adoption model. *Communication Theory, 13*(4), 345–365. doi:10.1111/j.1468–2885.2003.tb00296.x

Lin, T. C., & Sun, H. L. (2011). Connection as a form of resisting control: Foreign domestic workers' mobile phone use in Singapore. *Media Asia, 37*(4), 183–192.

Lindlof, T. R., & Taylor, B. C. (2011). *Qualitative communication research methods*. Thousand Oaks, CA: Sage.

Lindsley, S. L. (1999). Communication and "the Mexican way": Stability and trust as core symbols in Maquiladoras. *Western Journal of Communication, 63*(1), 1–31.

Ling, R. (2004). *The mobile connection: The cell phone's impact on society*. San Francisco, CA: Morgan Kaufmann.

Ling, R. (2007). Children, youth and mobile communication. *Journal of Children and Media, 1*(1), 60–67.

Ling, R. S. (2008). *New tech, new ties: How mobile communication is reshaping social cohesion*. Cambridge, MA: MIT Press.

Ling, R. S. (2012). *Taken for grantedness: The embedding of mobile communication into society*. Cambridge, MA: MIT Press.

Ling, R., Parekh, P., Zainudeen, A., & Galpaya, H. (2017 (Online first)). Rationalization of mobile telephony by small-scale entrepreneurs in Myanmar. *Information Communication & Society*.

Ling, R., & Yttri, B. (2002). Hyper-coordination via mobile phones in Norway. In J. Katz & M. Aakhus (Eds.), *Perpetual contact: Mobile communication, private talk, public performance* (pp. 139–169). Cambridge: Cambridge University Press.

Liu, H. (2014). Beyond co-ethnicity: The politics of differentiating and integrating new immigrants in Singapore. *Ethnic and Racial Studies, 37*(7), 1225–1238. doi: 10.1080/01419870.2014.892630

Liu, S. (2007). Living with others: Mapping the routes to acculturation in a multicultural society. *International Journal of Intercultural Relations, 31*(6), 761–778.

Lockhart, B. M. (Ed.). (2010). *Studies in Malaysian and Singapore history*. Kuala Lumpur: Royal Asiatic Society.

Lu, Y., Samaratunge, R., & Härtel, C. E. J. (2012). The relationship between acculturation strategy and job satisfaction for professional Chinese immigrants in the Australian workplace. *International Journal of Intercultural Relations, 36*(5), 669–681. doi:10.1016/j.ijintrel.2012.04.003

Lu, Y., Samaratunge, R., & Härtel, C. E. J. (2013). Acculturation attitudes and affective workgroup commitment: Evidence from professional Chinese immigrants in the Australian workplace. *Asian Ethnicity, 14*(2), 206–228. doi:10.1080/14631369.2012.722445

Lyon, D. (2008). Biometrics, identification and surveillance. *Bioethics, 22*(9), 499–508.

Mackay, H., & Gillespie, G. (1992). Extending the social shaping of technology approach: Ideology and appropriation. *Social Studies of Science, 22*(4), 685–716. doi:10.2307/285460

MacKenzie, D. A., & Wajcman, J. (1999). *The social shaping of technology* (2nd ed.). Buckingham, Philadelphia: Open University Press.

Madianou, M., & Miller, D. (2011). Mobile phone parenting: Reconfiguring relationships between Filipina migrant mothers and their left-behind children. *New Media & Society, 13*(3), 457–470. doi:10.1177/1461444810393903

Mahler, S. J. (2001). Transnational relationships: The struggle to communicate across borders. *Identities, 7*(4), 583–619. doi:10.1080/1070289X.2001.9962679

Manase, I. (2013). Zimbabwean migrants in the United Kingdom, the new media and networks of survival during the period 2000–2007. *African Identities, 11*(1), 33–46. doi:10.1080/14725843.2013.775839

Mante-Meijer, E., & H. van de Loo. (1998). Blurring of the life spheres: Flexibility and teleworking. In A. Kant & E. Mante-Meijer, (Eds.), *Blurring boundaries: When are Information and Communication Technologies coming home*. Stockholm: Telia/COST 248.

Mantovani, G. (2012). New media, migrations and culture: From multi- to interculture. In L. Fortunati, R. Pertierra, & J. Vincent (Eds.), *Migration, diaspora, and information technology in global societies* (pp. 21–34). New York: Routledge.

Marden, C. F., & Meyer, G. (1968). *Minorities in American society* (3rd ed.). New York: Van Nostrand.

Marger, M. N. (2001). The use of social and human capital among Canadian business immigrants. *Journal of Ethnic and Migration Studies, 27*(3), 439–453. doi:10.1080/136918301200266167

Marranci, G. (2011). Integration, minorities and the rhetoric of civilization: The case of British Pakistani Muslims in the UK and Malay Muslims in Singapore. *Ethnic and Racial Studies, 34*(5), 814–832. doi:10.1080/01419870.2010.539702

Massey, D. S. (1988). Economic development and international migration in comparative perspective. *Population and Development Review, 14*(3), 383–413.

Mathieu, J. E., & Zajac, D. M. (1990). A review and meta-analysis of the antecedents, correlates, and consequences of organizational commitment. *Psychological Bulletin, 108*(2), 171–194.

McGregor, J., & Pasura, D. (2010). Diasporic repositioning and the politics of re-engagement: Developmentalising Zimbabwe's diaspora?. *Round Table, 99*(411), 687–703. doi:10.1080/00358533.2010.530413

McKeown, A. (2004). Global migration, 1846–1940. *Journal of World History, 15*(2), 155–189.

McMillan, A., & Lopez, T. B. (2001). Socialization and acculturation: Organizational and individual strategies toward achieving P-O fit in a culturally diverse society. *The Mid-Atlantic Journal of Business, 37*(1), 19–34.

Mendola, M. (2008). Migration and technological change in rural households: Complements or substitutes? *Journal of Development Economics, 85*(1–2), 150–175. doi:10.1016/j.jdeveco.2006.07.003

Mendoza, R. H. (1984). Acculturation and sociocultural variability. In J. L. Martinez & R. H. Mendoza (Eds.), *Chicano psychology* (pp. 61–75). Orlando, FL: Academic Press.

Metykova, M. (2010). Only a mouse click away from home: Transnational practices of Eastern European migrants in the United Kingdom. *Social Identities, 16*(3), 325–338. doi:10.1080/13504630.2010.482418

Meyer, J. P., & Allen, N. J. (1991). A three-component conceptualization of organizational commitment. *Human Resource Management Review, 1*(1), 61–89.

MHA. (2014). Report of the committee of inquiry into the Little India riot on 8 December 2013: Ministry of Home Affairs, Singapore. Retrieved from https://www.mha.gov.sg/Newsroom/press-releases/PublishingImages/Pages/Report-of-the-Committee-of-Inquiry-into-the-Little-India-Riot-on-8-December-2013/Little%20India%20Riot%20COI%20report%20-%202014–06–27.pdf

Miller, M. J., Yang, M., Lim, R. H., Hui, K., Choi, N.-Y., Fan, X., . . . Blackmon, S. k. (2013). A test of the domain-specific acculturation strategy hypothesis. *Cultural Diversity and Ethnic Minority Psychology, 19*(1), 1–12. doi:10.1037/a0030499

Mishra, K., Boynton, L., & Mishra, A. (2014). Driving employee engagement: The expanded role of internal communications. *Journal of Business Communication, 51*(2), 183–202. doi:10.1177/2329488414525399

MoM. (2017a). *Foreign workforce numbers*. Ministry of Manpower, Government of Singapore. Retrieved from http://www.mom.gov.sg/documents-and-publications/foreign-workforce-numbers

MoM. (2017b). *Various types of housing and their specific requirements*. Ministry of Manpower, Government of Singapore. Retrieved from http://www.mom.gov

.sg/passes-and-permits/work-permit-for-foreign-worker/housing/various-types-of-housing

Mon, M. (2010). Burmese labour migration into Thailand: Governance of migration and labour rights. *Journal of the Asia Pacific Economy, 15*(1), 33–44.

Mosse, D., Gupta, S., Mehta, M., Shah, V., Rees, J., & Team, KRIBP. (2002). Brokered livelihoods: Debt, labour migration and development in tribal Western India. *Journal of Development Studies, 38*(5), 59–88.

Mowday, R. T., Steers, R. M., & Porter, L. W. (1979). The measurement of organizational commitment. *Journal of Vocational Behavior, 14*(2), 224–247. doi:10.1016/0001–8791(79)90072–1

Muchinsky, P. M. (1977). Organizational communication: Relationships to organizational climate and job satisfaction. *Academy of Management Journal, 20*(4), 592–607.

Mumford, L. (1934). *Technics and civilization*. New York: Harcourt, Brace & Co.

Munck, R., Schierup, C. U., & Wise, R. D. (2013). *Migration, work and citizenship in the new global order*. London: Taylor & Francis.

Murray, K. E., & Marx, D. M. (2013). Attitudes toward unauthorized immigrants, authorized immigrants, and refugees. *Cultural Diversity and Ethnic Minority Psychology, 19*(3), 332–341. doi:10.1037/a0030812

Nah, A. M. (2012). Globalisation, sovereignty and immigration control: The hierarchy of rights for migrant workers in Malaysia. *Asian Journal of Social Science, 40*(4), 486–508. doi:10.1163/15685314–12341244

Naravane, V. (2013, October 24). The war on migration can't be won. *The Hindu*. Retrieved from http://www.thehindu.com/opinion/interview/the-war-on-migration-cant-be-won/article5265658.ece

Navas, M., García, M. C., Sánchez, J., Rojas, A. J., Pumares, P., & Fernández, J. S. (2005). Relative Acculturation Extended Model (RAEM): New contributions with regard to the study of acculturation. *International Journal of Intercultural Relations, 29*(1), 21–37.

Nazroo, J., & Karlsen, S. (2003). Patterns of identity among ethnic minority people: Diversity and commonality. *Ethnic and Racial Studies, 26*(5), 902–930. doi:10.1080/0141987032000109087

Nedelcu, M. (2012). [SPECIAL ISSUE] Migrants' new transnational habitus: Rethinking migration through a cosmopolitan lens in the digital age. *Journal of Ethnic and Migration Studies, 38*(9), 1339–1356. doi:10.1080/1369183x.2012.698203

Ngan, R., & Ma, S. (2008). The relationship of mobile telephony to job mobility in China's Pearl River Delta. *Knowledge, Technology & Policy, 21*(2), 55–63.

Ntale, M. E., & Duma, S. E. (2011). The costs and benefits of nurse migration on families: A Lesotho experience. *Curationis, 34*(1), 1–8.

Ogburn, W. F. (1964). *On culture and social change: Selected papers*. Chicago: University of Chicago Press.

Ono, H., & Zavodny, M. (2008). Immigrants, English Ability and the Digital Divide. *Social Forces, 86*(4), 1455–1479.

O'Reilly, C. A., & Chatman, J. (1986). Organizational commitment and psychological attachment: The effects of compliance, identification, and internalization on prosocial behavior. *Journal of Applied Psychology, 71*(3), 492–499. doi:10.1037/0021–9010.71.3.492

O'Reilly, C. A., III., Chatman, J., & Caldwell, D. F. (1991). People and organizational culture: A profile comparison approach to assessing person-organization fit. *Academy of Management Journal, 34*(3), 487–516.

Overseas Filipinos. (2017). *2016 Survey on Overseas Filipinos*. Retrieved from https://psa.gov.ph/content/2016-survey-overseas-filipinos

Parham, A. A. (2004). Diaspora, community and communication: Internet use in transnational Haiti. *Global Networks, 4*(2), 199–217.

Park, R. E. (1928). Human migration and the marginal man. *American Journal of Sociology, 33*(6), 881–893. doi:10.2307/2765982

Park, R. E. (1971). *The immigrant press and its control*. New York: Harper & Row.

Pasek, J., More, E., & Romer, D. (2009). Realizing the social internet? Online social networking meets offline civic engagement. *Journal of Information Technology & Politics*, 6, 197–215.

Peng, Y., & Choi, S. Y. P. (2013). Mobile phone use among migrant factory workers in South China: Technologies of power and resistance. *The China Quarterly, 215*, 553–571. doi:10.1017/S0305741013000738

Perez, Y. V., & Ben-David, Y. (2012). Internet as freedom—does the internet enhance the freedoms people enjoy? *Information Technology for Development, 18*(4), 293–310. doi:10.1080/02681102.2011.643203

Pérez-Armendáriz, C., & Crow, D. (2010). Do migrants remit democracy? International migration, political beliefs, and behavior in Mexico. *Comparative Political Studies, 43*(1), 119–148. doi:10.1177/0010414009331733

Perlmann, J., & Waldinger, R. (1997). Second generation decline? Children of immigrants, past and present—a reconsideration. *International Migration Review, 31*(4), 893–922. doi:10.2307/2547418

Philippine Statistics Authority. (2017). *Philippine international migration data*. Retrieved from https://unstats.un.org/unsd/demographic-social/meetings/2017/bangkok—international-migration-data/Session%203/Session%203%20Philippines.pdf

Phinney, J. S., Horenczyk, G., Liebkind, K., & Vedder, P. (2001). Ethnic identity, immigration, and well-being: An interactional perspective. *Journal of Social Issues, 57*(3), 493–510.

PIB. (2013). *India is the second-largest mobile phone user in world.* Press Information Bureau, Government of India. Retrieved from http://pib.nic.in/newsite/erelease.aspx?relid=85669

Piontkowski, U., Florack, A., Hoelker, P., & Obdrzálek, P. (2000). Predicting acculturation attitudes of dominant and non-dominant groups. *International Journal of Intercultural Relations, 24*(1), 1–26.

Piper, N. (2005). *Country study: Singapore*. Report submitted to Friedrich Ebert Stiftung under Project on Migrant Labor in Southeast Asia. Retrieved from http://www.fes.de/aktuell/focus_interkulturelles/focus_1/documents/8_000.pdf

Population White Paper. (2013). *A sustainable population for a dynamic Singapore: Population White Paper*. National Population and Talent Division. Retrieved from

https://www.nptd.gov.sg/PORTALS/0/HOMEPAGE/HIGHLIGHTS/population-white-paper.pdf
Portes, A. (1999). Conclusion: Towards a new world—the origins and effects of transnational activities. *Ethnic and Racial Studies, 22*(2), 463–477. doi:10.1080/014198799329567
Portes, A., & Bach, R. L. (1985). *Latin journey: Cuban and Mexican immigrants in the United States*. Berkeley: University of California Press.
Portes, A., & Sensenbrenner, J. (1993). Embeddedness and immigration: Notes on the social determinants of economic action. *American Journal of Sociology, 98*(6), 1320–1350.
Portes, A., & Zhou, M. (1993). The new second generation: Segmented assimilation and its variants. *Annals of the American Academy of Political and Social Science, 530*(1), 74–96. doi:10.1177/0002716293530001006
PTI. (2014, January 1). Riots in Little India inexcusable: Singapore PM, *The Hindu*. Retrieved from http://www.thehindu.com/news/international/south-asia/riots-in-little-india-inexcusable-singapore-pm/article5525505.ece
Qiu, J. L. (2009). *Working-class network society: Communication technology and the information have-less in urban China*. Cambridge, MA: MIT Press.
Qiu, L. (2016). Sanders is right: Migrant laborers in Malaysia are forced into working under "slave-like conditions." *Politifact*. Retrieved from http://www.politifact.com/global-news/statements/2016/apr/12/bernie-s/sanders-right-migrant-laborers-malaysia-are-forced/
Qureshi, S. (2012). As the global digital divide narrows, who is being left behind? *Information Technology for Development, 18*(4), 277–280. doi:10.1080/02681102.2012.730656
Rahman, M. M. (2017). *Bangladeshi migration to Singapore: A process-oriented approach*. Singapore: Springer Nature.
Rahman, M. M., Yong, T. T., & Ullah, A. K. M. A. (Eds.). (2014). *Migrant remittances in South Asia: Social, economic and political implications*. London: Palgrave Macmillan.
Rahman, M. M., & Kiong, T. C. (2013). Integration policy in Singapore: A transnational inclusion approach. *Asian Ethnicity, 14*(1), 80–98. doi:10.1080/14631369.2012.710403
Rahman, M. M., & Lian, K. F. (2005). Bangladeshi migrant workers in Singapore: The view from the inside. *Asia-Pacific Population Journal, 20*, 63–88.
Raijman, R. (2013). Foreigners and outsiders: Exclusionist attitudes towards labour migrants in Israel. *International Migration, 51*(1), 136–151. doi:10.1111/j.1468-2435.2011.00719.x
Rath, J., & Kloosterman, R. (2000). Outsiders' business: A critical review of research on immigrant entrepreneurship. *International Migration Review, 34*(3), 657–681. doi:10.2307/2675940
Redfield, R., Linton, R., & Herskovits, M. (1936). Memorandum for the study of acculturation. *American Anthropologist, 38*, 149–152.
Reid, F. J. M., & Reid, D. J. (2010). The expressive and conversational affordances of mobile messaging. *Behaviour & Information Technology, 29*(1), 3–22. doi:10.1080/01449290701497079

Reis, R. (2010). Uses of mass media for adaptation purposes: A quantitative study of Brazilian immigrants in Los Angeles. *Journal of Intercultural Communication*, (24). Retrieved from http://www.immi.se/intercultural/nr24/reis.htm

Riedel, J., Wiesmann, U., & Hannich, H. J. (2011). An integrative theoretical framework of acculturation and salutogenesis. *International Review of Psychiatry, 23*(6), 555–564.

Riketta, M., & Van Dick, R. (2005). Foci of attachment in organizations: A meta-analytic comparison of the strength and correlates of workgroup versus organizational identification and commitment. *Journal of Vocational Behavior, 67*(3), 490–510. doi:10.1016/j.jvb.2004.06.001

Riordan, C. M., & Shore, L. M. (1997). Demographic diversity and employee attitudes: An empirical examination of relational demography within work units. *Journal of Applied Psychology, 82*(3), 342–358.

Roberts, B. R., & Portes, A. (2006). Coping with the free market city: Collective action in six Latin American cities at the end of the twentieth century. *Latin American Research Review, 41*(2), 57–83. doi:10.2307/3874669

Rogers, E. M. (2003). *Diffusion of innovations* (5th ed.). New York: Free Press.

Roldan, G. Z. (2009, May). Mobile phones and migrant workers in Malaysia. Paper presented at the "Mobile 2.0: Beyond Voice?" Pre-conference workshop at the International Communication Association (ICA) Conference, Chicago, Illinois. Retrieved from http://www.lirneasia.net/wp-content/uploads/2009/05/final-paper_roldan.pdf

Rubdy, R., & McKay, S. L. (2013). "Foreign workers" in Singapore: Conflicting discourses, language politics and the negotiation of immigrant identities. *International Journal of the Sociology of Language, 2013*(222), 157–185. doi:10.1515/ijsl-2013–0036

Rudmin, F. W. (2003). Critical history of the acculturation psychology of assimilation, separation, integration, and marginalization. *Review of General Psychology, 7*(1), 3–37.

Rudmin, F. W., & Ahmadzadeh, V. (2001). Psychometric critique of acculturation psychology: The case of Iranian migrants in Norway. *Scandinavian Journal of Psychology, 42*(1), 41–56.

Rudnick, A. (2009). *Working gendered boundaries: Temporary migration experiences of Bangladeshi women in the Malaysian export industry from a multi-sited perspective*. Amsterdam: The University Press.

Rumbaut, R. G. (2008). Reaping what you sow: Immigration, youth, and reactive ethnicity. *Applied Developmental Science, 12*(2), 108–111. doi:10.1080/10888690801997341

Rupert, J., Jehn, K. A., Engen, M. L. v., & Reuver, R. S. M. d. (2010). Commitment of cultural minorities in organizations: Effects of leadership and pressure to conform. *Journal of Business and Psychology, 25*(1), 25–37. doi:10.2307/40605743

Ryan, L. (2010). Becoming Polish in London: Negotiating ethnicity through migration. *Social Identities, 16*(3), 359–376. doi:10.1080/13504630.2010.482425

Ryan, L., Sales, R., Tilki, M., & Siara, B. (2008). Social networks, social support and social capital: The experiences of recent Polish migrants in London. *Sociology, 42*(4), 672–690.

Ryder, A. G., Alden, L. E., & Paulhus, D. L. (2000). Is acculturation unidimensional or bidimensional?: A head-to-head comparison in the prediction of demographics, personality, self-identity, and adjustment. *Journal of Personality and Social Psychology, 79*, 49–65. doi:10.1037/0022–3514.79.1.49

Sahoo, A. K. (2016). Diaspora, transnationalism and development. In S. Irudaya Rajan (Ed.), *India migration report 2014: Diaspora and development* (pp. 47–59). London: Routledge.

Saks, A. M. (2006). Antecedents and consequences of employee engagement. *Journal of Managerial Psychology, 21*(7), 600–619. doi:10.1108/02683940610690169

Salomon, G. (1985). The study of television in a cross-cultural context. *Journal of Cross-Cultural Psychology, 16*(3), 381–397. doi:10.1177/0022002185016003007

Sam, D. L. (2000). Psychological adaptation of adolescents with immigrant backgrounds. *Journal of Social Psychology, 140*(1), 5–25. doi:10.1080/00224540009600442

Sánchez, P., & Salazar, M. (2012). Transnational computer use in urban Latino immigrant communities. *Urban Education, 47*(1), 90–116. doi:doi:10.1177/0042085911427740

Sandhu, K. S. (1993). Indian immigration and settlement in Singapore. In K. S. M. Sandhu (Ed.), *Indian communities in Southeast Asia* (pp. 774–787). Singapore: Times Academic Press and Institute of Southeast Asian Studies.

Sarker, R. (2016). Migration and employment: A study of Bangladeshi male migrant workers in Malaysia. In K. F. Lian, M. M. Rahman, & Y. bin Alas (Eds.), *International migration in Southeast Asia: Continuities and discontinuities* (pp. 125–148). Singapore: Universiti Brunei Darussalam, Institute of Asian Studies; Springer.

Saunders, R. J., Warford, J. J., & Wellenius, B. (1994). *Telecommunications and economic development*. Baltimore, MD: John Hopkins University Press

Sayegh, L., & Lasry, J. C. (1993). Immigrants adaptation in Canada: Assimilation, acculturation, and orthogonal cultural identification. *Canadian Psychology, 34*(1), 98–109.

Schmalzbauer, L. (2004). Searching for wages and mothering from afar: The case of Honduran transnational families. *Journal of Marriage and Family, 66*(5), 1317–1331.

Schmalzbauer, L. (2008). Family divided: The class formation of Honduran transnational families. *Global Networks, 8*(3), 329–346.

Schwartz, S. J., Unger, J. B., Zamboanga, B. L., & Szapocznik, J. (2010). Rethinking the concept of acculturation: Implications for theory and research. *American Psychologist, 65*(4), 237–251. doi:10.1037/a0019330

Sey, A. (2007, May). *What have mobile phones wrought? Innovative calling practices to manage cost*. Paper presented at the 57th Annual Conference of the International Communication Association, San Francisco, CA.

Sey, A. (2011). "We use it different, different": Making sense of trends in mobile phone use in Ghana. *New Media & Society, 13*(3), 375–390. doi:10.1177/1461444810393907

Sharp, L. (1952). Steel axes for stone-age Australians. *Human Organization*, 17–22.

Shibutani, T., & Kwan, L. (1965). *Ethnic stratification: A comparative approach*. New York: Macmillan.

Shrum, L. J. (2002). Media consumption and perceptions of social reality: Effects and underlying processes. In J. Bryant & D. Zillmann (Eds.), *Media effects: Advances in theory and research* (pp. 50–73). Mahwah, NJ: Erlbaum.

Sibal, H. T., & Foo, S. (2016). A study on the information seeking behaviour of Singapore-based Filipino domestic workers. *Information Development, 32*(5), 1570–1584. doi:10.1177/0266666915615929

Siddique, H. (2012, November 28). Singapore's first strike in 25 years shines spotlight on racial tensions. *The Guardian*. Retrieved from https://www.theguardian.com/world/2012/nov/28/chinese-bus-drivers-strike-singapore

Siddiqui, T. (2005). International migration as a livelihood strategy of the poor. In T. Siddiqui (Ed.), *Migration and development: Pro-poor policy choices*. Bangladesh: The University Press.

Siew, S. C. (1986). *Problems faced by foreign workers in Singapore construction industry* (Unpublished honors thesis). Department of Building and Estate Management, National University of Singapore, Singapore.

Signorielli, N., & Morgan, M. (Eds.). (1990). *Cultivation analysis: New directions in media effects research*. Newbury Park, CA: Sage.

Silverstone R., Hirsch E., & Morley D. (1992). Information and communication technologies and the moral economy of the household. In R. Silverstone & E. Hirsch (Eds.), *Consuming technologies: Media and information in domestic spaces* (pp. 15–31). New York: Routledge.

Silvey, R. (2007). Unequal borders: Indonesian transnational migrants at immigration control. *Geopolitics, 12*(2), 265–279. doi:10.1080/14650040601168917

Sing Stat. (2017). *Population trends 2017.* Department of Statistics Singapore. Retrieved from http://www.singstat.gov.sg/docs/default-source/default-document-library/publications/publications_and_papers/population_and_population_structure/population2017.pdf

Singapore's angry migrant workers. (2013, December 27). *The New York Times*. Retrieved from http://www.nytimes.com/2013/12/28/opinion/singapores-angry-migrant-workers.html?_r=1&

Skuza, J. (2007). Humanizing the understanding of the acculturation experience with phenomenology. *Human Studies, 30*(4), 447–465. doi:10.1007/s10746–007–9073–6

Sniderman, P. M., Hagendoorn, L., & Prior, M. (2004). Predisposing factors and situational triggers: Exclusionary reactions to immigrant minorities. *The American Political Science Review, 98*(1), 35–49. doi:10.2307/4145295

Solomon, J. (2016). *A subaltern history of the Indian diaspora in Singapore: Gradual disappearance of untouchability 1872–1965*. New York: Routledge.

Souter, D., Scott, N., Garforth, C., Jain, R., Mascararenhas, O., & McKemey, K. (2005). *The economic impact of telecommunications on rural livelihoods and poverty reduction: A study of rural communities in India (Gujarat), Mozambique, and Tanzania*. Commonwealth Telecommunications Organisation for UK Department for International Development. Retrieved from https://www.gov.uk/dfid-research-outputs

Spencer-Rodgers, J., & McGovern, T. (2002). Attitudes toward the culturally different: The role of intercultural communication barriers, affective responses, consensual stereotypes, and perceived threat. *International Journal of Intercultural Relations, 26*(6), 609–631.

Spindler, G. C., & Goldschmidt, W. (1952). Experimental design in the study of cultural change. *Southwestern Journal of Anthropology, 8*, 68–83.

Stark, O., & Bloom, D. E. (1985). The new economics of labor migration. *American Economic Review, 75*(2), 173–178.

Stephan, W. G., Renfro, L. C., Esses, V. M., Stephan, C. W., & Martin, T. (2005). The effects of feeling threatened on attitudes toward immigrants. *International Journal of Intercultural Relations, 29*(1), 1–19.

Stephens, K. K., & Ford, J. L. (2014, May). *Banning mobile devices: Workplace policies that selectively exclude can shape crisis communication.* Paper presented at the 11th International ISCRAM Conference, University Park, Pennsylvania, US.

Stephens, K., Zhu, Y., Harrison, M., Iyer, M., Hairston, T., & Luk, J. (2017). Bring Your Own Mobile Device (BYOD) to the hospital: Layered boundary barriers and divergent boundary management strategies. In *Proceedings of the 50th Hawaii International Conference on System Sciences*. Retrieved from http://scholarspace.manoa.hawaii.edu/handle/10125/41584

Stonequist, E. V. (1935). The problem of the marginal man. *American Journal of Sociology, 41*(1), 1–12. doi:10.2307/2768176

Subervi-Velez, F. A. (1986). The mass media and ethnic assimilation and pluralism: A review and research proposal with special focus on Hispanics. *Communication Research, 13*(1), 71–96. doi:10.1177/009365028601300105

Sunoo, D. H., Trotter, E. P., & Aames, R. L. (1980). Media use and learning of English by immigrants. *Journalism Quarterly, 57*(2), 330–333.

Sunstein, C. R. (2001). *Republic.com*. Princeton, NJ: Princeton University Press.

Suroor, H. (2011, August 22). Indian expatriates take out anti-corruption rallies in Europe. *The Hindu*. Retrieved from http://www.thehindu.com/news/national/article2384499.ece

Tadmor, C. T., Tetlock, P. E., & Peng, K. (2009). Acculturation strategies and integrative complexity: The cognitive implications of biculturalism. *Journal of Cross-Cultural Psychology, 40*(1), 105–139.

Tazanu, P. M. (2015). On the liveness of mobile phone mediation: Youth expectations of remittances and narratives of discontent in the Cameroonian transnational family. *Mobile Media & Communication, 3*(1), 20–35. doi:10.1177/2050157914545801

"Telecom subscriber base in India grows to 1.18 billion in Feb: Trai." (2017). *Times of India*. Retrieved from http://timesofindia.indiatimes.com/business/india-business/telecom-subscriber-base-in-india-grows-to-1–18-billion-in-feb-trai/articleshow/58443836.cms

Teo, Y., & Piper, N. (2009). Foreigners in our homes: Linking migration and family policies in Singapore. *Population, Space and Place, 15*(2), 147–159. doi:10.1002/psp.545

Thomas, M., & Lim, S. S. (2011). ICT use and female migrant workers in Singapore. In J. E. Katz (Ed.), *Mobile communication: Dimensions of social policy* (pp. 175–189). New Brunswick, NJ: Transaction.

Thomas, W. I., & Znaniecki, F. (1958). *The Polish peasant in Europe and America*. New York: Dover Publications.

Thompson, E. C. (2009). Mobile phones, communities and social networks among foreign workers in Singapore. *Global Networks, 9*(3), 359–380.

Toffler, A. (1981). *The third wave*. New York: Bantam Books.

Toolan, M. (1997). What is critical discourse analysis and why are people saying such terrible things about it? *Language and Literature, 6*(2), 83–103. doi:10.1177/096394709700600201

"Trouble in Little India." (2013, December 14). *The Economist*. Retrieved from https://www.economist.com/news/asia/21591626-nearly-unbelievable-full-scale-riot-obedient-city-state-trouble-little-india

Tsai, J. L., Ying, Y. W., & Lee, P. A. (2000). The meaning of "being Chinese" and "being American": Variation among Chinese American young adults. *Journal of Cross-Cultural Psychology, 31*, 302–332.

Tsui, A. S., Egan, T. D., & O'Reilly, C. A. (1992). Being different: Relational demography and organizational attachment. *Administrative Science Quarterly, 37*(4), 549–579.

"2G network to be completely shut by Apr 18." (2017). Channel News Asia. Retrieved from http://www.channelnewsasia.com/news/singapore/2g-network-to-be-completely-shut-by-apr-18–8596618

Tynes, R. (2007). Nation-building and the diaspora on Leonenet: A case of Sierra Leone in cyberspace. *New Media & Society, 9*(3), 497–518. doi:10.1177/1461444807076980

UNHCR. (2015). *Bay of Bengal and Andaman sea: Proposals for action*. Retrieved from http://www.unhcr.org/protection/migration/55682d3b6/action-plan-2015-bay-bengal-andaman-sea.html

UNICEF. (2013). *Migration profiles*. Retrieved from https://esa.un.org/miggmg profiles/indicators/files/Malaysia.pdf

United Nations. (2014). We can end poverty: Millennium development goals and beyond 2015. Retrieved from http://www.un.org/millenniumgoals/bkgd.shtml

United Nations. (2015a). *International migration report 2015*. Department of Economic & Social Affairs. Retrieved from http://www.un.org/en/development/desa/population/migration/publications/migrationreport/docs/MigrationReport2015.pdf

United Nations. (2015b). *International migration stock 2015*. Retrieved from http://www.un.org/en/development/desa/population/migration/data/estimates2/estimates15.shtml

United Nations. (2017). Department of Economic and Social Affairs. Population Division. *Trends in international migrant stock: The 2017 revision*. Retrieved from United Nations database, POP/DB/MIG/Stock/Rev.2017.

Vacaru, M. A., Shepherd, R. M., & Sheridan, J. (2014). New Zealand Youth and Their Relationships with Mobile Phone Technology. *International Journal of Mental Health and Addiction, 12*(5), 572–584. doi:10.1007/s11469–014–9488-z

van de Vijver, F. J. R., Breugelmans, S. M., & Schalk-Soekar, S. R. G. (2008). Multiculturalism: Construct validity and stability. *International Journal of Intercultural Relations, 32*(2), 93–104. doi:10.1016/j.ijintrel.2007.11.001

van de Vijver, F. J. R., Helms-Lorenz, M., & Feltzer, M. J. A. (1999). Cognitive assessment in education in a multicultural society. *International Journal of Psychology*, 34, 149–162.
Vancea, M., & Olivera, N. (2013). E-migrant women in Catalonia: Mobile phone use and maintenance of family relationships. *Gender, Technology and Development, 17*(2), 179–203. doi:10.1177/0971852413488715
Vanden Abeele, M., Antheunis, M. L., & Schouten, A. P. (2014). Me, myself and my mobile: A segmentation of youths based on their attitudes towards the mobile phone as a status instrument. *Telematics and Informatics, 31*(2), 194–208. doi:10.1016/j.tele.2013.04.004
Vedder, P., & Virta, E. (2005). Language, ethnic identity and the adaptation of Turkish immigrant youth in the Netherlands and Sweden. *International Journal of Intercultural Relations, 29.* 317–337.
Velayutham, S., & Wise, A. (2005). Moral economies of a translocal village: Obligation and shame among south Indian transnational migrants. *Global Networks, 5*(1), 27–47.
Vertovec, S. (2004). Cheap calls: The social glue of migrant transnationalism. *Global Networks, 4*(2), 219–224.
Viswanath, K., & Arora, P. (2000). Ethnic media in the United States: An essay on their role in integration, assimilation, and social control. *Mass Communication and Society, 3*(1), 39–56.
Volkmer, I. (2014). *The global public sphere*. Cambridge: Polity.
Wallis, C. (2008). Techno-mobility and translocal migration: Mobile phone use among female migrant workers in Beijing. In M. I. Srinivasan & V. V. Ramani (Eds.), *Gender Digital Divide* (pp. 196–216). Hyderabad: Icfai University Press.
Wallis, C. (2013). *Technomobility in China: Young migrant women and mobile phones*. New York: New York University Press.
Ward, C. (2001). The A, B, Cs of acculturation. In D. R. Matsumoto (Ed.), *The handbook of culture & psychology* (pp. 411–446). New York: Oxford University Press.
Ward, C., & Kennedy, A. (1994). Acculturation strategies, psychological adjustment, and sociocultural competence during cross-cultural transitions. *International Journal of Intercultural Relations, 18*(3), 329–343. doi:10.1016/0147–1767(94)90036–1
Ward, C., & Kennedy, A. (1999). The measurement of sociocultural adaptation. *International Journal of Intercultural Relations, 23*(4), 659–677. doi:10.1016/S0147–1767(99)00014–0
Ward, C., & Rana-Deuba, A. (1999). Acculturation and adaptation revisited. *Journal of Cross-Cultural Psychology, 30*(4), 422–442. doi:10.1177/0022022199030004003
Warschauer, M. (2002). Reconceptualizing the Digital Divide. *First Monday*, 7(7).
Waters, M. C., Tran, V. C., Kasinitz, P., & Mollenkopf, J. H. (2010). Segmented assimilation revisited: Types of acculturation and socioeconomic mobility in young adulthood. *Ethnic and Racial Studies, 33*(7), 1168–1193. doi:10.1080/01419871003624076
Weber, M. (1904/1949a). Objectivity in social science and social policy. In E. A. Shils & H. A. Finch (Eds.), *The methodology of social sciences* (pp. 49–112). Chicago, IL: Free Press.

Weber, M. (1904/1949b). The meaning of "ethical neutrality" in Sociology and Economics. In E. A. Shils & H. A. Finch (Eds.), *The methodology of social sciences* (pp. 1–47). Chicago, IL: Free Press.

Wei, R. (2008). Motivations for using the mobile phone for mass communications and entertainment. *Telematics and Informatics, 25*(1), 36–46. doi:10.1016/j.tele.2006.03.001

Wilding, R. (2006). "Virtual" intimacies? Families communicating across transnational contexts. *Global Networks, 6*(2), 125–142. doi:10.1111/j.1471–0374.2006.00137.x

Williams, R. (1974). *Television, technology and cultural form.* London: Fontana.

Wilson, J. (2013). *Exploring the past, present and future of cultural competency research: The revision and expansion of the sociocultural adaptation construct.* Unpublished doctoral dissertation. Victoria University of Wellington. Retrieved from http://www.victoria.ac.nz/cacr/research/mirips/socio-cultural-adaptation-scale

Wilson, K. L., & Portes, A. (1980). Immigrant enclaves: An analysis of the labor market experiences of Cubans in Miami. *American Journal of Sociology, 86*(2), 295–319.

Wimmer, A., & Glick Schiller, N. (2002). Methodological nationalism and beyond: Nation-state building, migration and the social sciences. *Global Networks, 2*(4), 301–334.

Winner, L. (1980). Do artifacts have politics? *Daedalus*, 121–136.

Winner, L. (1993). Upon opening the black box and finding it empty: Social constructivism and the philosophy of technology. *Science, Technology, & Human Values, 18*(3), 362–378. doi:10.2307/689726

Wirth, W., von Pape, T., & Karnowski, V. (2008). An integrative model of mobile phone appropriation. *Journal of Computer-Mediated Communication, 13*(3), 593–617.

Woolgar, S. (1991). The turn to technology in social studies of science. *Science, Technology, & Human Values, 16*(1), 20–50. doi:10.2307/690038

World Bank. (2016a). Migration and remittances factbook 2016. Retrieved from https://openknowledge.worldbank.org/bitstream/handle/10986/23743/9781464803192.pdf?sequence=3&isAllowed=y

World Bank. (2016b). *GDP per capita (current US$).* Retrieved from http://data.worldbank.org/indicator/NY.GDP.PCAP.CD

World Bank. (2016c). Personal remittances, received (% of GDP). Retrieved from http://data.worldbank.org/indicator/BX.TRF.PWKR.DT.GD.ZS

World Bank. (2017). *Migration and remittances: Recent developments and outlook.* Retrieved from http://pubdocs.worldbank.org/en/992371492706371662/MigrationandDevelopmentBrief27.pdf

Xiang, B., & Lindquist, J. (2014). Migration infrastructure. *International Migration Review, 48*(s1), S122-S148. doi:10.1111/imre.12141

Yang, H., Yang, P., & Zhan, S. (2017). Immigration, population, and foreign workforce in Singapore: An overview of trends, policies, and issues. *HSSE, 6*(1), 10–25.

Yang, K. (2008). A preliminary study on the use of mobile phones amongst migrant workers in Beijing. *Knowledge, Technology & Policy, 21*(2), 65–72. doi:10.1007/s12130–008–9047–7

Yaple, P., & Korzenny, F. (1989). Electronic mass media effects across cultures. In M. K. Asante, W. B. Gudykunst, & E. Newmark (Eds.), *Handbook of international and intercultural communication* (pp. 295–317). Newbury Park, CA: Sage Publications.

Yeh, C. J., & Inose, M. (2003). International students' reported English fluency, social support satisfaction, and social connectedness as predictors of acculturative stress. *Counseling Psychology Quarterly, 16*, 15–28.

Yeoh, B. S. A., & Huang, S. (2010). Transnational domestic workers and the negotiation of mobility and work practices in Singapore's home spaces. *Mobilities, 5*(2), 219–236. doi:10.1080/17450101003665036

Yeoh, B. S. A., Huang, S., & Gonzalez III, J. (1999). Migrant female domestic workers: Debating the economic, social and political impacts in Singapore. *International Migration Review, 33*(1), 114–136.

Yi, C.-C. (2013). Changing East Asian families: Values and behaviors. *International Sociology, 28*(3), 253–256. doi:10.1177/0268580913488397

Yum, J. O. (1982). Communication diversity and information acquisition among Korean immigrants in Hawaii. *Human Communication Research, 8*(2), 154–169.

Zachariah, K. C., & Rajan, I. S. (2009). *Migration and development: The Kerala Experience*. Delhi: Daanish Books.

Zachariah, K. C., & Rajan, I. S. (2016). Kerala migration study 2014. *Economic and Political Weekly, 51*(6), 66–71.

Zhou, M. (2004). Revisiting ethnic entrepreneurship: Convergencies, controversies, and conceptual advancements. *International Migration Review, 38*(3), 1040–1074.

Zubrzycki, J. (1958). The role of the foreign-language press in migrant integration. *Population Studies, 12*(1), 73–82.

Index

Page references for figures are italicized

About the Authors

Rajiv George Aricat (PhD, 2016, Nanyang Technological University, Communication Studies) is a Research Fellow at the School of Social Sciences, NTU, Singapore. His research areas include adoption and use of Information Communication Technology (ICT) among low-income groups, technopolitics of new media and immigrant integration. For his doctoral research he studied the mobile phone use of South Asian migrant workers in Singapore. He has worked in research projects on mobile communication in Myanmar, and rumors in social media. His papers have appeared in *International Journal of Communication, Journal of Information Technology & Politics,* and *Information Technology & People*.

Rich Ling (PhD, University of Colorado, sociology) is the Shaw Foundation Professor of Media Technology at Nanyang Technological University, Singapore, and has an adjunct position at the University of Michigan. For the past two decades, Ling has studied the social consequences of mobile communication. He has written *The Mobile Connection* (Morgan Kaufmann, 2004), *New Tech, New Ties* (MIT, 2008) and most recently *Taken for Grantedness* (MIT, 2012). He is the editor-in-chief of the *Journal of Computer-Mediated Communication* and is a founding co-editor of *Mobile Media and Communication* (Sage) and the Oxford University Press Series Studies in Mobile Communication. In 2016, he was named a fellow of the International Communication Association.

www.ingramcontent.com/pod-product-compliance
Ingram Content Group UK Ltd.
Pitfield, Milton Keynes, MK11 3LW, UK
UKHW021906150926
475560UK00003B/205

9 781498 552509